Colloid and Surface Chemistry

A Laboratory Guide for Exploration of the Nano World

Colloid and Surface Chemistry

A Laboratory Guide for Exploration of the Nano World

Seyda Bucak | Deniz Rende

CRC Press
Taylor & Francis Group
Boca Raton London New York

CRC Press is an imprint of the
Taylor & Francis Group, an **informa** business

CRC Press
Taylor & Francis Group
6000 Broken Sound Parkway NW, Suite 300
Boca Raton, FL 33487-2742

First issued in paperback 2019

© 2014 by Taylor & Francis Group, LLC
CRC Press is an imprint of Taylor & Francis Group, an Informa business

No claim to original U.S. Government works

ISBN-13: 978-1-4665-5310-1 (hbk)
ISBN-13: 978-0-367-37901-8 (pbk)

Library of Congress Cataloging-in-Publication Data

Bucak, Seyda.
    Colloid and surface chemistry : a laboratory guide for exploration of the nano world / Seyda Bucak, Deniz Rende.
        pages cm
    "A CRC title."
    Includes bibliographical references and index.
    ISBN 978-1-4665-5310-1 (pbk. : alk. paper)
    1. Surface chemistry--Laboratory manuals. I. Rende, Deniz. II. Title.

QD506.B73 2014
541'.345078--dc23   2013045724

Visit the Taylor & Francis Web site at
http://www.taylorandfrancis.com

and the CRC Press Web site at
http://www.crcpress.com

We dedicate this book to our life partners, who have been there for us

Contents

List of figures

List of tables

Preface

The principles of colloid and interface science are shaping today's most advanced technologies from nanomedicine to electronic nanorobots. This area, although traditionally considered a part of physical chemistry, currently involves more widespread applications than ever due to its relevance to environmental and biological problems. Through new advances, colloid and interface science has become a truly interdisciplinary subject integrating chemistry, physics, and biology.

This laboratory book is designed to help students to understand the basic principles of colloid and interface science through experiments underlining the fundamental principles. We aimed to introduce these concepts to junior- and senior-level undergraduate students, from various disciplines, such as chemistry, chemical engineering, materials science and engineering, as well as biological sciences and engineering, who are introduced to the subject matter for the first time. The objective of this book is to bridge the gap between the theory behind the colloids and surface chemistry and applications in the field.

We have been guided by several publications in this field, all of which have offered generous resources as textbooks. Two of these, *Introduction to Modern Colloid Science* (Hunter, 1994) and *Introduction to Colloid and Surface Chemistry* (Shaw, 1992), convey the fundamentals of the topic with basic descriptions. These books provide the principles without exemplifying the concepts, since the principles are presented in a traditional way. *Principles of Colloid and Surface Chemistry* (Hiemenz, 1985) is one of the fundamental books in the field, including the derivations of the mathematical expressions and providing the extensive theoretical information. *The Colloidal Domain: Where Physics, Chemistry, Biology, and Technology Meet* (Advances in Interfacial Engineering) (Evans and Wennerström, 1999) presents a straightforward approach in handling colloidal behavior, guides the reader with a summary of concepts, and provides a comprehensive understanding through theory and examples. These concepts are also illustrated by the relevant experimental techniques.

Surface and Colloid Chemistry: Principles and Applications (Birdi, 2009) explains the principles with many daily life examples, including cleaners and cosmetics, and demonstrates how surface and colloid chemistry has become a part of natural events and industrial processes. This comprehensive book presents a practical perspective to the subject; we intended to supplement this broad information with laboratory practices. *Colloid Science: Principles, Methods and Applications* (Cosgrove, 2010) presents industrial applications in the field, including food, pharmaceutical, agrochemical, cosmetics, polymer, paint, and oil industries. It explains the formation, control, and characterization of the colloids to the graduate-level students and postgraduate chemists. In this book, we aimed to present the concepts through experiments such that principles and laboratory techniques are learned as fundamental research tools and the applications are seamlessly integrated with theory.

Chapter 1 starts with the aspects of research methodology in science and exemplifies the landmarks of designing successful experiments. This chapter is in accordance with the ethics in science, which we believe is an inseparable component of scientific research. This chapter also contains practical information on the importance of data collection, analysis of data, laboratory bookkeeping, and writing laboratory reports.

Chapter 2 reviews the techniques that are frequently used in the characterization and analysis of colloidal structures to perform successful experiments. In this chapter, we aimed to introduce the working principles of characterization techniques, including surface tension measurements, viscosity and rheological measurements, electrokinetic techniques, scattering and diffraction techniques, as well as microscopy. Each of these sections was written by a distinguished researcher in the corresponding field. We would like to express our sincere gratitude to those who contributed to this chapter: Patrick Underhill from Rensselaer Polytechnic Institute, Marek Kosmulski from Lublin University of Technology, Ulf Olsson from Lund University, and Nico Sommerdijk from Eindhoven Technical University.

In this book, 19 experiments were conducted: each has been designed to convey a concise understanding of the corresponding topic. We conceive the experiments from a student's perspective. Therefore, each of the experiments in Chapters 3–5 starts with a purpose of the experiment, which simply identifies the problem and continues with the background information about the topic. This section is supplemented with essential studies in the field for those who want to have a deeper understanding of the topic. The pre-laboratory questions aim to induce critical thinking about the experiment even before coming to the laboratory. The step-by-step procedures are followed by tables and instructions, which allow the students to record their data and observations as well as guides the students in the calculations. The post-laboratory questions highlight the essential parts of the experiment and facilitate further discussion.

While keeping in mind that laboratory experiments should be modular, include interchangeable components, such as chemicals or measurements, and provide a time frame of the experiment, we have added notes to the instructor section at the end of each experiment. These tips and suggestions would be valuable for the instructor while designing the experiments throughout the semester.

One other aspect of this book is the possibility of integrating the experiments. For instance, according to the flow of the course, 1 week could be allocated to synthesize nanoparticles, followed by a concept experiment. We believe this organization would provide the instructor flexibility as well as leverage the ability of the students to integrate the concepts and remind that these experiments are the components of a broad research field.

Chapter 3 contains 11 experiments about colloids and surfaces. This chapter starts with a sedimentation experiment, which emphasizes the importance of particle size and medium in the sedimentation process. This experiment is followed by critical micelle concentration determination of various surfactants and measurement of the surface tension of a nonionic surfactant to derive the Gibbs adsorption isotherm. The dynamic surface tension measurements in this experiment allow to identifying the adsorption mechanism. The relationship between the contact angle and the wettability of solid surfaces is explored in terms of surface roughness, and Zisman plot is used to determine the critical surface tension. The conformations of polymers in different solvents are explored by measuring viscosity and using Mark–Houwink plot. The electrostatic interactions between oppositely charged particles are studied by the addition of polyelectrolytes to silica nanoparticle surface by tracking the surface charges. The stability of colloidal particles is very important in applications, and hence this phenomenon is demonstrated with three experiments: altering the stability of gold nanoparticles by the addition of different electrolytes, introducing emulsions, and pickering emulsions by examining relationship between wetting of particles and emulsion stabilization, and foam stability of different surfactants. Depending on the structure, colloids can be used in different applications; to have a control over the final structure, the understanding of phase diagrams is of great importance. In this respect, the miniemulsions, which serve as nanosized reactors for polymerization reactions, are prepared with phase inversion.

Chapter 4 covers the different techniques of preparing nanoparticles. The first three experiments in the chapter explain straightforward protocols to synthesize silver nanoparticles with a reduction reaction, magnetic nanoparticles with coprecipitation reaction, and silica nanoparticles with a hydrolysis condensation reaction. An example of reduction in a confined environment is provided with an experiment to synthesize nickel nanoparticles. The synthesis of latex particles in a miniemulsion experiment is another synthesis in a confined environment, offering an application to understand the miniemulsion systems.

Chapter 5 shows how theory turns into practice and contains experiments of general applications, which students come across every day. The first experiment in this chapter explains the preparation of mayonnaise from a colloidal science perspective. Body wash and cream formulation experiments intend to introduce different types of colloids and how these formulations change according to the final product. The experiments in this chapter explain the theoretical perspective of the daily life products.

Among our collaborators, friends, and students who contributed to this book, we would like to emphasize the contribution of Nihat Baysal and Cem Levent Altan, for their participation at each level and writing fundamentals in scientific computing section and microscopy, respectively. We are indebted to Rahmi Ozisik for his suggestions about diffraction section and guidance in some experiments. Cuneyt Tas and Daniel Lewis provided valuable comments and suggestions in the diffraction section.

Binnaz Coskunkan, Elcin Yenigul Tulunay, Ecegul Tok, Merve Seyhan, Sukriye Sulamaz, Umit Ecem Yarar, Bekir Cakici, each performed excellent in repeating the experiments over and over and helping us to revise the procedures. We are indebted to Naz Zeynep Atay, Erde Can, Gulengul Duman, Ryan Gilbert, Anne Hynes, Mustafa Ozilgen, Sibel Ozilgen, Lennart Piculell, Sevinc Rende, David Steytler, and Georgi G. Yordanov for providing feedback on various sections of this book.

Authors

Seyda Bucak has more than 15 years of experience in chemistry laboratory practices, predominantly in colloids and surface chemistry. During her PhD at the University of East Anglia, UK, and postdoctoral studies at the Massachusetts Institute of Technology, USA, she conducted experiments in a variety of areas within the field, from liposomes to metallic nanoparticles, exploring the fundamentals and applications of colloidal materials. She has numerous publications on self-assembly, colloidal material, and nanoparticles. She has also coauthored *General Chemistry Laboratory Book* and *Physical Chemistry Laboratory Book* published by Yeditepe University Press. She has been working as an associate professor in the Department of Chemical Engineering at Yeditepe University since 2011. She is involved in teaching colloid and surface chemistry at the undergraduate and graduate levels. Currently, her main research areas are in the synthesis and applications of magnetic nanomaterials and peptide self-assembly.

Deniz Rende has nearly 10 years of experience in undergraduate-level chemical engineering laboratory courses. She actively participated in establishing and coordinating various undergraduate laboratories to be used in teaching general chemistry, physical chemistry, and unit operations courses in the Department of Chemical Engineering, Yeditepe University, Istanbul, Turkey. After receiving her PhD from the Department of Chemical Engineering, Bogazici University, she worked as a postdoctoral research associate at Rensselaer Nanotechnology Center, Rensselaer Polytechnic Institute, where she conducted research on supercritical fluid-assisted processing of polymer nanocomposites to microcellular structures, viscoelastic properties of polymer nanocomposites, and magnetic nanoparticles. She is currently teaching as an adjunct professor and appointed as Proximal Probe Laboratory Manager in the Department of Materials Science and Engineering, Rensselaer Polytechnic Institute. Her current research involves viability and reactivity of cells in response to activated magnetic nanoparticles.

Contributors

Cem Levent Altan
Technische Universiteit Eindhoven
Eindhoven, the Netherlands

and

Yeditepe University
Istanbul, Turkey

Nihat Baysal
Rensselaer Polytechnic Institute
Troy, New York

Marek Kosmulski
Lublin University of Technology
Lublin, Poland

and

Leibniz-Institut für
Polymerforschung
Dresden, Germany

Ulf Olsson
Lund University
Lund, Sweden

Nico A.J.M. Sommerdijk
Technische Universiteit Eindhoven
Eindhoven, the Netherlands

Patrick Underhill
Rensselaer Polytechnic Institute
Troy, New York

Symbols

Symbol		Chapter
h	Height	35
H	Magnetic strength	198
h, k, l	Miller indices	70
[i]	Luminous Intensity	9
I, [I]	Electric current	9, 64
I	Scattering intensity	81
ΔI_{ref}	Excess scattered intensity of the reference solvent	94
$\vec{k}$	Incident wave vector	81
$\vec{k}_0$	Wave vector	81
k_B	Boltzmann constant	97, 143
l_{max}	Extended length	177
l_{tail}	Hydrophobic tail length	210
L, l, [L]	Length	9, 28, 43
L	Wetted length	36
L_e	Entrance length	48
m, [M]	Mass	9, 32
M	Molarity	135
M	Molecular weight	158
M	Magnetic moment	197
n	Number of moles	10
n	Refractive index of the solvent	95
n_{ref}	Refractive index of the reference solvent	94
[N]	Amount of a substance	9
N	Number of experiments	19
p	Momentum	84
P, p, [P]	Pressure	9, 10, 29, 43
$P(q)$	Particle formfactor	87
q	Scattering angle	81
Q	Volumetric flow rate	43
r_{core}	Droplet size	210
$\Delta R(q)$	Excess Raleigh ratio	94
R	Universal gas constant	10
R, r	Radius	30, 43, 87, 125
R	Electrical resistance	65
Re	Reynolds number	48
R_g	Radius of gyration	150
R_H, $r_{hydrodynamic}$	Hydrodynamic radius	97, 210

Symbol		Chapter
R_{ref}	Raleigh ratio of the reference solvent	94
s	Standard deviation	19
S	Cross sectional area	64
$S(q)$	Structure factor	88
t, [t]	Time	9, 128
t_{obs}	Observation time	41
T, [T]	Temperature	9, 10, 29, 97
T	Torque	45
u	Internal energy	10
U	Mean velocity	43
U	Electric potential	65
v, [v]	Velocity	9, 10, 44
v_s	Settling velocity	126
V, [V]	Volume	9, 10
V_m	Volume of the hydrophobic surfactant region	210
V_p	Volume of particles	89
V_s	Sample volume	86
V_s	Volume of a sphere	87
V_T	Total potential energy	162
w	Weight	16
w	Work	27, 132
w_0	Water-to-surfactant molar ratio	210
x	Measured value	18
$\bar{x}$	Arithmetic average of the measurements	19
z	Elevation	10
α, β, γ	Interaxial angles between planes	69
γ	Surface tension	27, 132
$\dot{\gamma}_a$	Apparent shear rate	44
$\dot{\gamma}_R$	Shear rate at the wall	44
Γ	Relaxation rate	97
Γ	Surface concentration	140
$\delta(x)$	Dirac delta function	84
Δ	Difference	129
ε	Dielectric constant	59
ε	Molar extinction coefficient	203
ζ	Zeta-potential	49

Symbol		Chapter
$[\eta]$	Intrinsic viscosity	152
η	Solvent viscosity	98
η	Kinematic viscosity	152
η_{inh}	Inherent viscosity	153
η_r	Relative viscosity	153
η_{red}	Reduced viscosity	153
η_{sp}	Specific viscosity	153
θ	Contact angle	35, 145
Θ	Angle of the space between the cone and plate	45
κ	Parameter close to one	47
κ	Reciprocal Debye length	60
λ	Relaxation time	41
λ	Wavelength	68, 128
μ	True value	18
μ	Viscosity	43
μ	Electrophoretic mobility	56
μ	Magnetic permeability	197
$\rho, [\rho]$	Density	9, 34, 43, 127
ρ	Resistivity	65
ρ	Scattering power	83
ρ_f	Fluid density	127
ρ_p	Particle density	127
σ	Standard deviation	91
τ_r	Stress at the wall	43
ϕ	Volume fractions	89
χ	Magnetic susceptibility	197
ω, Ω	Angular velocity	34, 44, 125

chapter one

Scientific research

The research process

Research is primarily a quest for knowledge and understanding. It is a useful, interesting experience, and an intriguing technique to improve the quality of human life. Scientific research is a long process consisting of several steps: observation, hypothesis, experimentation, and interpretation. Each step is important and determines the course of research in its own way.

Observations serve as the starting point for throwing questions and are fundamental to the designing of experiments. Hence, observation leads to selection and ultimately to description (Wilson, 1952). Observation comes from consciousness and curiosity and can be improved through critical reading and the desire to integrate knowledge (Smith, 1998). A research topic aims to identify and fill a gap in the literature, and hence it starts with an overview of the present and past literature in a specific area (Greenfield, 1996). One of the most important steps in observation is the immediate recording of the data. The second step involves the question of bias, which reflects an individual's existing prejudices. The conditions should be arranged in such a way that the observer's bias on events should not interfere with the observation itself. Numerical observations are always featured, not only to quantify the relations among the observations, but also to compare independent observations with each other (Wilson, 1952). A good observation guides one to questions to identify relationships.

The proposed explanation of an observed relationship is called a hypothesis. It originates from observations, aiming to explain the relation between two phenomena through variables. Hypotheses were thought to be perfect and sufficient enough for gaining knowledge in earlier ages, but now they are regarded as important guides that can be supported only by experimental evidence. Hypotheses originate in literature studies and should be used in the design of relevant experiments. Before formulating a hypothesis, a detailed and careful examination of the literature is required in order to refrain oneself from performing experiments conducted previously. Hence, a thorough literature reading will facilitate further refinement of the question and help one to develop a finalized hypothesis (Smith, 1998). Developing a testable, novel hypothesis in the confines of one's scientific expertise for many is the most personal and creative part of the research process. However, for others with less experience, hypothesis development may be the most challenging aspect of the research process.

The core of scientific research is experimentation. Experimentation leads to the acquisition of data. The data generated from one particular experiment are commonly quantified through mathematical formulae to display outcomes that are compared to other experiments. As Pythagoras suggested: "Mathematics is the way to understand the universe.... Numbers are the measure of all things" (Hamming, 1980).

Good experimentation begins with a methodical planning and performance. The planning should be in accordance with the developed hypothesis and should identify the factors or variables to be tested (with the appropriate inclusion of controls), produce experimental outcomes worthy of analysis, stringent methods of analysis, and subsequent statistical testing of the results. Control groups are almost the same as the experimental groups, except that with the experimental groups a test is performed to prove or disprove the hypothesis. In certain situations, standards are used to compare experimental data to known and proven results. For instance, calorimetric heat measurements generated through the combustion of organic compounds are made relative to the standard samples of benzoic acid, which is prepared in the same manner. Hence, since the combustion experiment is performed and the same process used within the experiments was used to assess combustion of the benzoic acid sample, the benzoic acid results should agree with others' measurements of benzoic acid combustion, ensuring that the specific sample measurement is valid (Wilson, 1952). The fundamentals of a successful experiment design are discussed in the forthcoming sections.

Interpretation is the analysis of the results collected after running the experiments and explaining what the results mean through writing and/or display of graphs and/or images. One key aspect of interpreting data is that the data collected in a series of experiments are consistent with the underlying research question. As a part of the interpretation process, statistical analyses are included, the results of which are presented in the form of tables or plots and the findings are discussed and compared with other investigator's findings. Statistical analysis determines which experimental groups are different from each other and which experimental groups are different from the controls. Representation of the data, in tables or graphs, compares and highlights differences in the data.

Ethics in science

Ethics in research refers to individual and communal codes of behavior based on a set of principles for conducting research. In this sense, scientific ethics could be defined as the standards of conduct for scientists in their professional endeavors, and ethics are considered to be within a wide range of activities from technological developments to environmental issues to the study of living organisms. Research ethics is important not only to prevent harm to the public and research subjects, but also to protect the environment.

Although a complete discussion of ethics in science is beyond the scope of this book, this section intends to stimulate the thinking of ethics in scientific studies and demonstrate how ethics is an integral component of scientific research. Self-regulation is one of the key parameters in identifying and controlling errors to maintain integrity. The scientific community examines and assesses the results of a research; depending on the credibility of the findings, has right to discuss and present the correct information to others as an ordinary part of scientific inquiry.

Scientific ethics is significant mainly in three distinct areas: (i) scientific work, which involves performing the study itself and documentation of the research; (ii) the interactions within the scientific community in terms of intellectual property, collaborations, and contributions to an original research; and (iii) the needs and values of the society.

Validity of research and documentation

The process of making and reporting scientific work has a strong ethical component. Accuracy of the results from a scientific work relies on correct experimentation, data recording and interpretation, as well as reproducibility. Scientific work should be transparent and well documented in all aspects, starting from recording each and every step in an experimental work with the observations. The work is explained in sufficient clarity to prevent any information loss in time. Assumptions should also be well documented.

Interactions with the scientific community

The ideas and work of a scientist is the intellectual property of that scientist or the employer where the scientist resides. Scientific work is cumulative; it progresses in small increments that build over prior work (Bauer, 1992). Therefore, it is important to acknowledge the work of others contributing to the study. Arguably, a property implies the use of a specific entity; however, it is possible to distinguish intellectual property as an intangible property. Tangible properties are more easily understood as a concept. For example, a private property, such as a land or a house, is a physical entity one can own. On the other hand, a property, such as a street, is owned by public. These public properties are to be used by all citizens. In both cases, either a private owner or the public maintains the control over the property. However, when it comes to intangible property, ownership is more abstract. Information, which is an intangible property, should be accessible to anyone but the control of its dissemination belongs to the owner (Lenk, 2007). Examples of disregarding intellectual property is the minimization of the significance of previous work, demonstration of others' ideas as self-ideas, and omission of prior research and presenting without citing these studies.

One essential component of scientific achievement is the collaboration among scientists. One should include anyone as an author who has directly contributed to a scientific research, without considering the amount of work or time s/he spent on it. There is a thin line between the co-authorship and acknowledgement, which distinguishes itself from the original contribution of ideas and work (co-authorship) and allowing access to some facilities, but not the idea or work (acknowledgement). Hence co-authorship may include conducting experiments, analysis of experimental data, and interpretation of the results.

Intellectual property, citing others' works, collaborations, and individual contribution to a scientific research determine how interactions among the scientific community evolve around research ethics.

The needs and values of society

All scientists have a duty to ensure that their work ultimately serves socially desirable ends. Without ethical scrutiny, the scientists could forget their responsibilities and track into private fields of focus on special rather than social benefits. It is the scientists' responsibility to ensure that their research does not cause unjustified risks to other individuals and animals, thereby jeopardizing environmental welfare. Research fields that jeopardize the environment are mainly based on the unsustainable products and processes that directly harm the environment, cause environmental pollution, and reduce the availability of sustainable products (Shrader-Frechette, 1994).

When a research is conducted, it is expected to offer benefits to the society. A majority of the scientific studies receive monetary support from various institutions, such as government or private institutions. These programs are valuable in terms of identifying directly the needs of the society; however, a scientist should refrain himself from conducting biased research, which is directed to produce specific results, and convert any public interest to a private profit. In addition, from another perspective, the scientists' responsibility in a sponsored research project is not to waste the resources for a specific project.

These primary ethical considerations should always be borne in mind while conducting research.

Design of experiments

Before planning any experiment, the researcher should have a very good understanding of the nature and the underlying principles of the topic. It is essential to design the experiments based on a hypothesis. This can be achieved only by completely understanding the theory behind the problem (Wilson, 1952). The design of experiments is an invaluable tool for identifying critical parameters, optimizing the processes, and getting the most with minimum cost. A good experimental design includes the minimum number of experiments to obtain the broadest and valuable information. Hence, it should enable the researcher to successfully distinguish the effects of parameters in a process and to reduce the possible noise in the system. Experimental designs include the steps to ensure that

1. Enough number of measurements and repetitions is made
2. Random control experiments are performed
3. The variability in the data is maintained

In addition to these, other considerations should also be taken into account, such as time, cost, and effective use of resources. These factors are also a part of the decision process. Notably, sufficient number of repetitions should be made to meet the objectives of the study and to prevent waste of resources and costs.

Specifically in chemical analysis, purity of the material produced at the end of a reaction should be controlled. For instance, representative samples from the entire material should be collected, and samples should be pooled to ensure that an average value is achieved. Samples should be randomized but individual characteristics should be retained. The methods should be compared during the analysis. To do so, the sample should be divided into two parts: analysis techniques should be applied to each portion and individual results should be compared. Corresponding random experiments should be carried out and compared with the experimental measurements. The factors to be assessed should be identified in designing the experiments to ensure that each factor is measured. The random experiments should also be designed for comparison; in this sense, time and cost are also considerably important.

Randomization of samples is required to remove bias in the experiments. In experiments, the order of measurements depends on the choice of the experimenter. However, the order of measurements should be randomized to prevent any bias. The design of random experiments is mainly straightforward: the samples are labeled by a randomly generated sequence regardless of the duplicates in the sample. A random sequence can be generated from many resources, including calculators and computer software.

Data collection is the most important part of an experiment, since poor data collection may affect the overall analysis result. For some cases, if the data collected are at a low resolution, the overall results might not be representative of the experiment. For instance, consider recording the absorbance of a solution in a UV spectrophotometer at a low resolution, 0.7 ABS, rather than 0.683 ABS. If the primary question is to determine the presence of a material, the former is sufficient to indicate this information. However, if the major concern is the concentration of a material in a solution where even a 0.1% concentration change is important, then high-resolution values should be recorded.

To start a successful experimental design, a clear identification of the objective is needed. The objective should be translated into precise questions. It is always useful to work in teams, as well as involving experts from other fields. The out-of-box views improve the experimental design quality, for instance, consulting a statistician is always helpful in precisely identifying the experimental factors, in deciding the statistical methods by which the data will be analyzed, and in eliminating bias by creating random experiments.

One basic point is to list the materials, methods, measurements, and equipment to arrive at the desired outcome. When the final goal is identified, the factors and responses should be listed. The limits of each factor are also a determining factor. The number of measurements can be conceived by determining each factor's relation to the response. For instance, the responses could be the yield of the process and responsible for the purity of the final product in a chemical reaction. Factors might include the concentration and addition rates of the reactants, time taken to complete the reaction, and also the temperature of the environment. If the relation between the response and the factor is linear, then two measurements are enough to explain this relation; however, if the relation is nonlinear, then at least three measurements should be made to ensure that the relation is conceived. Then according to each relation between the factors and response, the measurement should be leveled to reach the final design. Although, as a basic point, this approach is sufficient, one should bear in mind that scientific research is iterative. Each measurement, that is, result, gives birth to another question to be answered.

Two-level design is the general approach applied to design experiments that is to assign a low and high level to each factor, which might be shown as –1 and +1, respectively (Figure 1.1). The design space for a two-factor experiment, assuming that the relationship of each factor to the response is linear, is a square. If the number of factors is 3, then this design space translates into a cube.

Consider a reaction between two reactants, where concentration is varied for reactant A and type is varied for reactant B, as B1 and B2. In

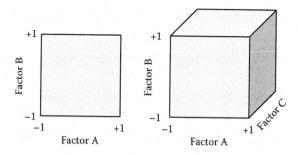

Figure 1.1 A two-level design. Design space for two and three factors, where the relationship to response is linear.

addition, temperature is controlled during the experiment (Table 1.1). The response of the reaction is measured by the purity of the final product. One possible design is as follows:

In a two-level design, these four factors will yield (2^3) eight experiments, and purity of the final product is recorded at each corresponding experiment (Table 1.2).

In terms of the purity of the final product, the effect of reactant A is determined by the average of response to each level tested, for instance, at low concentration of reactant A, the average purity is 65.58%, whereas at

Table 1.1 Example of Design Parameters in a Reaction Experiment

Factors	Low level	High level
Reactant A (concentration)	Low (−1)	High (+1)
Reactant B (type)	B1 (−1)	B2 (+1)
Temperature	Low (−1)	High (+1)

Table 1.2 The Outcomes of the Experiments by a Two-Level Experiment Design

	Reactant A (concentration)	Reactant B (type)	Temperature	Purity (%)
1	Low (−1)	B1 (−1)	Low (−1)	47.8
2	Low (−1)	B1 (−1)	High (+1)	94.1
3	Low (−1)	B2 (+1)	Low (−1)	66.2
4	Low (−1)	B2 (+1)	High (+1)	64.2
5	High (+1)	B1 (−1)	Low (−1)	44.6
6	High (+1)	B1 (−1)	High (+1)	48.9
7	High (+1)	B2 (+1)	Low (−1)	44.8
8	High (+1)	B2 (+1)	High (+1)	70.2

Table 1.3 Examples of Multiplicative Experimental Design to Assess Interactive Effects

	Reactant A	Reactant B	Temperature	AB	AT	BT	ABT	Purity (%)
1	−1	−1	−1	1	1	1	−1	47.8
2	−1	−1	1	1	−1	−1	1	94.1
3	−1	1	−1	−1	1	−1	1	56.2
4	−1	1	1	−1	−1	1	−1	64.2
5	1	−1	−1	−1	−1	1	1	74.6
6	1	−1	1	−1	1	−1	−1	78.9
7	1	1	−1	1	−1	−1	−1	74.8
8	1	1	1	1	1	1	1	70.2

high concentration, this value is 74.63%. The change in the response with respect to reactant A is a 9.05% increase in purity at high reactant A concentration. (The calculation of an average is explained in the "Statistical Significance" section.)

A multiplicative approach also shows interactive effects. For this approach, the −1 and +1 annotations are taken into account, where an interactive effect is the multiplication of these values. The average purity calculated according to the multiplication shows the interactive effect (Table 1.3).

The average response for a positive interaction (AB = 1) between A and B is 71.73%, whereas the negative interaction (AB = −1) yields 68.48% purity. This multidimensional approach could be extended for all levels.

Fundamentals of scientific computing

Nihat Baysal

Dimensions and units

Measurement plays a very important role in analyzing physical quantities. A dimension is a property that can be measured; length is used to understand the distance between the cafeteria and the classroom, time is used to find out how long does it take to walk to the classroom, and mass is used to express how heavy the books you carry are. Sometimes these base dimensions are combined to derive new types of measurements such as velocity (length/time), volume (length^3), and density (mass/volume). Both base and derived dimensions are listed in Table 1.4.

The derived dimensions listed in Table 1.4 can be expressed by using the base dimensions as follows:

Table 1.4 Base and Derived Dimension Symbols

Dimension	Symbol
Base	
Length	[L]
Mass	[M]
Time	[t]
Temperature	[T]
Electrical current	[I]
Luminous intensity	[i]
Amount of a substance	[N]
Derived	
Area	[A]
Volume	[V]
Velocity	[v]
Acceleration	[a]
Density	[ρ]
Force	[F]
Pressure	[P]
Energy	[E]

Area = Length × Length $\rightarrow$ $[A] = [L]^2$

Volume = Area × Length $\rightarrow$ $[V] = [A][L] = [L]^3$

$\text{Velocity} = \dfrac{\text{Length}}{\text{Time}}$ $\rightarrow$ $[v] = [L][t]^{-1}$

$\text{Accelaration} = \dfrac{\text{Velocity}}{\text{Time}}$ $\rightarrow$ $[a] = \dfrac{[L][t]^{-1}}{[t]} = [L][t]^{-2}$

$\text{Density} = \dfrac{\text{Mass}}{\text{Volume}}$ $\rightarrow$ $[\rho] = [M][L]^{-3}$

Force = Mass × Acceleration $\rightarrow$ $[F] = [M][L][t]^{-2}$

$\text{Pressure} = \dfrac{\text{Force}}{\text{Area}}$ $\rightarrow$ $[P] = \dfrac{[M][L][t]^{-2}}{[L]^2} = [M][L]^{-1}[t]^{-2}$

Energy = Force × Length $\rightarrow$ $[E] = [M][L][t]^{-2}[L] = [M][L]^2[t]^{-2}$

Scientists develop models to help understand the behavior of natural events as well as to solve engineering problems. These models use several equations, which consist of parameters and variables. In order to obtain a dimensionally consistent model, one should check the dimensions of the elements in each equation and be sure that both sides of the equation reduce to the same base dimensions. This type of consistency check is called dimensional analysis.

EXAMPLE

Check whether the total energy equation is dimensionally consistent. It is defined as the sum of the kinetic, potential, and internal energies:

$$E_{total} = \frac{1}{2}mv^2 + mgz + mu$$

where m is the mass [M], v the velocity [v], z the elevation [L], u the specific internal energy [E], and g the gravitational acceleration [a].

$$[E] = [M][L]^2[t]^{-2} + [M][L][t]^{-2}[L] + [M][E][M]^{-1}$$

$$[M][L]^2[t]^{-2} = [M][L]^2[t]^{-2} + [M][L][t]^{-2}[L] + [M][M][L]^2[t]^{-2}[M]^{-1}$$

$$[M][L]^2[t]^{-2} = [M][L]^2[t]^{-2}$$

Since both sides of the equation ended up with the same base dimensions, it is said to be dimensionally consistent.

EXAMPLE

Find the dimension of the universal gas constant, R in the ideal gas equation:

$$PV = nRT$$

The dimensions of each variable in both sides of the equation should be written as

$$([M][L]^{-1}[t]^{-2})([L]^3) = [N]R[T]$$

To obtain the dimension of R, multiply both sides by $[N]^{-1}[T]^{-1}$:

$$R = ([M][L]^2[t]^{-2})([N]^{-1}[T]^{-1})$$

All physical quantities have a numerical value and a unit, which is defined as the standard size subdivision of a dimension. In order to compare the magnitudes of dimensions, people need to use their values with corresponding units. The units are standardized for each base dimension according to SI (Systeme International d'Unites), known as metric system. Another system, which is known as the English or British unit system, is nowadays being less used since SI system is becoming very popular in many industrialized countries and fast becoming the standard unit system. The list of base and derived units, which are used to quantify dimensions, is given in Tables 1.5–1.7.

In general practice, the numerical values of physical quantities range between 0.1 and 1000 in a standard decimal form. In order to express values falling out of this range, a prefix should be used. A prefix is a letter that is used in front of a unit and denotes a power-of-ten multiple of its value. For example, "kilogram" is used to state relatively heavier substances, or "millisecond" is used to express very short time intervals. The standard prefixes for SI units are listed in Table 1.8.

Now, it is time to learn basic rules of using SI units properly. Here is a brief list of rules, and Table 1.9 presents some common mistakes committed for the SI units:

1. Do not pluralize unit symbols. Otherwise, it may be confused with the unit "second (s)."

Table 1.5 Base SI Units and Associated Dimensions

Unit name	Symbol	Dimension
Meter	m	Length
Kilogram	kg	Mass
Second	s	Time
Kelvin	K	Temperature
Ampere	A	Electrical current
Candela	cd	Luminous intensity
Mole	mol	Amount of a substance

Table 1.6 Derived SI Units with a Specific Name and Associated Derived Dimensions

Unit name	Symbol	Base units	Dimension
Newton	N	$kg \cdot m \cdot s^{-2}$	Force
Pascal	Pa	$kg \cdot m^{-1} \cdot s^{-2}$	Pressure, stress
Joule	J	$kg \cdot m^2 \cdot s^{-2}$	Energy, work, heat
Watt	W	$kg \cdot m^2 \cdot s^{-3}$	Power
Hertz	Hz	s^{-1}	Frequency
Coulomb	C	$A \cdot s$	Electric charge
Volt	V	$kg \cdot m^2 \cdot s^{-3} \cdot A^{-1}$	Electric potential
Ohm	Ω	$kg \cdot m^2 \cdot s^{-3} \cdot A^{-2}$	Electric resistance
Weber	Wb	$kg^{-1} \cdot m \cdot s^{-2} \cdot A^{-1}$	Magnetic flux
Katal	kat	$mol \cdot s^{-1}$	Catalytic activity

Table 1.7 SI Units of Other Derived Dimensions

Dimension	SI units
Acceleration	$m \cdot s^{-2}$
Area	m^2
Concentration	$mol \cdot m^{-3}$
Density	$kg \cdot m^{-3}$
Energy	$N \cdot m$
Entropy	$J \cdot K^{-1}$
Heat transfer rate	$J \cdot s^{-1}$
Mass flow rate	$kg \cdot s^{-1}$
Specific energy	$J \cdot kg^{-1}$
Surface tension	$N \cdot m^{-1}$
Thermal conductivity	$W \cdot m^{-1} \cdot K^{-1}$
Velocity	$m \cdot s^{-1}$
Viscosity, dynamic	$Pa \cdot s$
Viscosity, kinematic	$m^{-2} \cdot s^{-1}$
Volume	M^3
Volume flow rate	$m^3 \cdot s^{-1}$
Weight	N

2. Do not use a period (.) after a unit symbol unless the symbol is at the end of the sentence.
3. Do not use invented unit symbols. For example, (sec), (amp), or (cc) should not be used for the units of "second," "ampere," or "cubic centimeters," respectively.
4. Use always lowercase letters for a unit symbol, with three exceptions. The first exception is for the units named after people, such as newton

Table 1.8 Standard Prefixes for SI Units

Multiple	Prefix	Symbol
10^{12}	tera	T
10^9	giga	G
10^6	mega	M
10^3	kilo	k
10^{-1}	deci	d
10^{-2}	centi	c
10^{-3}	milli	m
10^{-6}	micro	μ
10^{-9}	nano	n
10^{-12}	pico	p

Table 1.9 Proper and Improper Ways of Using SI Units

Proper	Improper	Rule
1.1 cm	1.1 cms	1
1.82 N	1.82 N	2
2.4 A	2.4 amps	3
0.435 J	0.435 j	4
4 m·s	4 m s	5
1.14 GN	1.14 kMN	7
2.02 Pa·s	2.02 Pa · s	8
15 s	15 sec	3

(N), pascal (Pa), and watt (W). The second exception is the usage of prefixes, such as mega (M), giga (G), and tera (T). The third exception is a little bit different: liter. Although both uppercase or lowercase versions of the letter "L" can be used to express liter, the uppercase version should be used to avoid the confusion with the number "1".

5. Use half-high dot (·) or dash (-) to separate multiple unit symbols to avoid confusion with prefixes. For example, if the dot in the unit "meter-second" (m · s) is not used, the unit could be interpreted as "millisecond" (ms). Although space can be used as a separator, it is not preferred in general practice.

6. If a unit with a prefix has an exponential power, it applies to both the prefix and the unit symbol. For example, (ms^2) means "squared-millisecond," not "meter-second square (m · s^{-2} or m/s^2)." This is the same reason why we should be very careful while using separators in multiple units.

7. Do not combine prefixes. There is no such thing as "kilo-mega-joule" (kMJ). Simply replace it by "giga-joule" (GJ).

8. Always insert a space between the numerical values and the unit symbols. Also there should not be a space between the prefix and the unit symbol.

9. Use italic type for variable and quantity symbols, and roman type for unit symbols regardless of the typeface used in the surrounding text. Superscripts and subscripts should also be italic type if they represent the variables, quantities, or iterable numbers. Otherwise, they should be in roman type. For example, c_p is used for specific heat capacity and T_i is used for the temperature of the inlet stream.

10. Do not mix any information with unit symbols or names. Use "the NaCl content is 10 g/L" instead of "10 g NaCl/L."

11. Do not mix unit symbols and unit names and do not apply mathematical operations to unit names. The usage of "kilogram/m^3" or "kg/cubicmeter" is improper, use "kg/m^3" instead.

Unit conversions

In most problems, the units of the variables do not match and need to be converted to a standard unit.

EXAMPLE

A school bus with a speedometer reading of 36.0 km/h passes through one of the author's house at 07:00 am. The distance between the house and the school is 660 m. Determine the time required for the bus to arrive at the school.

1. Convert the speed to [m/s]
   To find the time needed to reach the school from the house, the speed (velocity) of the bus should be converted to [m/s]:

$$\frac{36.0}{1} \times \frac{1000 \text{ m}}{1} \times \frac{1}{3600 \text{ s}} = \frac{36000}{3600} = 10.0 \text{ m/s}$$

2. Calculate the time required
   The distance should now be divided by the speed. The result gives the time in seconds:

$$\frac{660}{10.0/\text{s}} = 66.0 \text{ s}$$

EXAMPLE

Calculate the mass and weight of a stainless steel rectangular solid bar with the dimensions of 60 cm × 12 cm × 15 cm. The density of the bar can be taken as 5.0×10^3 kg/m³.

1. Calculate the mass of a solid bar

$$\text{mass} = \text{density} \times \text{volume}$$

$$\text{mass} = 5.0 \times 10^3 \left(\frac{\text{kg}}{\text{m}^3} \right) \times (60 \times 12 \times 15) \text{cm}^3 \times \frac{1.0 \text{ m}^3}{1.0 \times 10^6 \text{ cm}^3}$$

$$\text{mass} = 5.4 \text{ kg}$$

2. Calculate the weight of solid bar

$$\text{weight} = \text{mass} \times \text{gravitational acceleration}$$

$$\text{weight} = 5.4 \text{ kg} \times 9.8 \text{ m/s}^2 = 53 \text{ N}$$

Significant figures

Whenever a scientist needs to use a number in a calculation, that number should be used with confidence and the result should be as accurate as possible. For example, try to read a burette volume, which is shown in Figure 1.2. It is very obvious that the reading should be between 40 and 41 mL. Since the meniscus is above the midpoint, we can accept it approximately as 41 mL with confidence. But someone might say the value is read as 40.8 mL and another might say 40.7 mL. In this case, the limits of the burette should be considered and the value of 41 mL should be accepted with confidence.

The significant figure or significant digit in a number is defined as a digit that designates the reliability of a value as a result of a measurement or a calculation. In the above example, only two digits are significant for the burette volume reading: 41 mL. As indicated here, the results, answers, or readings should always be expressed with confidence and the number of significant digits used in these results should be correct. More significant digits than necessary should not be used in answers to give an impression that the result is more accurate.

There are certain rules to understand how many significant figures a number has. The following list gives these rules in a categorized way with some examples in parentheses. The significant figures in those examples are underlined.

1. All nonzero digits are significant (<u>876</u>, <u>12.34</u>, 0.<u>261</u>). There are some exceptions, which will be explained below.
2. All zeroes after the significant figures in integer numbers are not significant unless a decimal point is used at the end of the number (<u>62</u>00, <u>256</u>000, <u>3490</u>., <u>169000</u>). The first two numbers can also be expressed in scientific notation as <u>6.2</u> × 10³ and <u>2.56</u> × 10⁵, respectively.

Figure 1.2 Significance while reading a burette.

3. All zeroes between the significant figures or after the decimal point that are not necessary are significant (2.6045, 10.256, 0.450, 3.0).
4. All zeroes before the significant values in numbers smaller than 1 are not significant because these numbers can easily be expressed in scientific notation (0.00627, 0.00000404, 0.010).

The rules for significant figures that are used in calculations are as follows:

1. In a multiplication or a division operation, the number of significant figures in the result (product or quotient) should contain the fewest significant figures in the numbers used in operation.
2. In an addition or subtraction operation, the result should show significant figures as seen in the least precise number used in calculation.

EXAMPLE

Two objects are put in a container. The masses of the two objects and the container are 20.5, 4.675, and 1.2 kg, respectively. Calculate the total weight of the system and express the result by using appropriate number of significant figures.

1. Find the total weight
   In order to find the total weight, the total mass of the system should be multiplied by the gravitational acceleration, $g = 9.81$ m/s^2. The weight of each item will be calculated by multiplying the mass with g:

$$w_1 = m_1 \times g = 20.5 \times 9.81 = 201.105 = 201 \text{ N}$$

Since both numbers have the same number of significant figures, the result should also have the same number of significant figures. Similarly,

$$w_2 = m_2 \times g = 4.675 \times 9.81 = 45.86175 = 45.9 \text{ N}$$

$$w_3 = m_3 \times g = 1.2 \times 9.81 = 11.772 = 12 \text{ N}$$

Now, it is time for addition according to Rule 2:

$$w_{total} = w_1 + w_2 + w_3 = 201 + 45.9 + 12 = 258.9 = 259 \text{ N}$$

The least precise numbers in this operation are 201 and 12, that is why the result is rounded to 259.

Errors in measurements

While making measurements in the laboratory, it is important to know how good the measurement is. Therefore, two concepts, accuracy and precision of the measurements, are of great importance during data collection. These two concepts can be explained by a throwing dart illustration.

Experimental research requires the perfect fulfillment of the following steps:

1. A theory to be tested
2. Design of the experiment to verify the theory
3. Accumulation of data
4. Analysis of data
5. Conclusions: Proving or disproving the theory

Statistical significance

Accuracy is an important parameter while analyzing experimental data, since it refers to the agreement between the experimental value and the true value. An accurate measurement is obtained if the observed or measured value is close to the true value (Figure 1.3).

An experimental researcher would repeat the same experiment several times to be sure of the result that he/she is getting. As a student, of course you will get a chance to do an experiment only once most of the time. That makes error analysis even more important. While performing an experiment, it is inevitable that there will be some mistakes done even by the most careful scientist. Sometimes, these may arise from some instrumental faults. As a result, you will get values that will not exactly match what you will find in the literature as "true value." These true values themselves are a result of a very high number of experiments. All values are analyzed carefully and averaged to obtain the literature "true

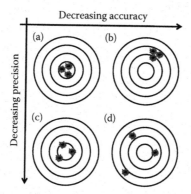

Figure 1.3 Illustration for precision and accuracy. (a) High accuracy, high precision, (b) low accuracy, high precision, (c) high accuracy, low precision, and (d) low accuracy, low precision.

value" of an unknown. Therefore, it is just normal that your one set of data will deviate from it. This deviation is called as error and it is defined with respect to a true value as

$$\text{error} = |x - \mu| \tag{1.1}$$

where x is the measured value and μ is the true value.

Errors can be represented as the percentage error, which is an absolute value of the error divided by the true value

$$\text{error}\% = \frac{|x - \mu|}{\mu} \times 100 \tag{1.2}$$

The errors in laboratory measurements can be categorized as systematic (determinate) and random (indeterminate).

Systematic errors arise from a bias that is placed on the measurement either by the instrument itself or by an improper method of reading or using the instrument. For example, an uncalibrated balance will read the mass either too high or too low all the time. Or, automatic pipettes with tips measure small volumes very accurately if the solution is aqueous. When organic solvents are used, especially those with high viscosity, the accuracy is less. This is an example of improper use of the instrument. The experimenter will always get a deviation in the same direction from the real value.

Random errors arise from intrinsic limitations in the sensitivity of the instrument or the experimenter him/herself. For example, a balance that reads only two decimal points may not be sensitive enough for a small mass or the volume from a pipette can be read incorrectly. Or, we may forget to control the temperature and fluctuations in the room temperature, which may cause a different reading each time. This type of errors does not take place in the same direction, meaning that they may be giving once a higher and then a lower reading. Given sufficient number of experiments, random errors, unlike systematic ones, tend to cancel each other out.

Simply, precision indicates degree of reproducibility of a measurement. A highly precise measurement is obtained when multiple experiments are performed, and at the end of each experiment similar values are collected, independent of the closeness to the true value (Figure 1.3). Precision is usually expressed in terms of the deviation of a set of results from the arithmetic mean of the set. It should always be borne in mind that good precision does not mean good accuracy.

As the number of repetitive measurements increased, the tendency of the measurements is generally obtained by calculating the arithmetic mean. The general definition of arithmetic mean is the summation of N numbers computed by some function and dividing by N:

$$\bar{x} = \frac{1}{N}\sum_{i=1}^{N} x_i \tag{1.3}$$

where x_i are the data recorded after each measurement, N the number of experiments performed, and $\bar{x}$ the arithmetic average of the measurements.

Another important issue in analyzing experimental data is the standard deviation, which provides a quantitative measure of the spread of the experimental data about the mean value:

$$s = \sqrt{\frac{1}{(N-1)}\sum_{i=1}^{N}(x_i - \bar{x})^2} \tag{1.4}$$

where x_i are the data recorded after each measurement, N the number of experiments performed, $\bar{x}$ the arithmetic average of the measurements, and s the standard deviation.

EXAMPLE

Two students measure the room temperature with an electronic thermometer and repeat their measurements five times. The room temperature was known to be 21.8°C. According to their measurements, which student's measurement is more precise and/or more accurate? Support your conclusion by calculating average, standard deviation, error, and percent error.

	T_1 (°C)	T_2 (°C)
1	20.5	21.9
2	20.2	22.4
3	21.1	21.3
4	19.9	22.6
5	20.4	21.2

1. Calculate the average and standard deviation of the measurements

$$\bar{T_1} = \frac{20.5 + 20.2 + 21.1 + 19.9 + 20.4}{5},$$

$$\bar{T_1} = 20.4°C,$$

$$s_1 = 0.444$$

Similarly,

$$\bar{T}_2 = 21.9°C,$$

$$s_2 = 0.630$$

The true value of temperature is 21.8°C, and the measurements recorded by the second student are more accurate. However, the standard deviation of the first student is less; hence these measurements are more precise.

2. Calculate error and percent error of the measurements

$$error_1 = |\bar{T}_1 - T_1| = 1.40°C$$

$$error_1\% = \frac{1.40°C}{21.8°C} \times 100 = 6.42\%$$

Similarly,

$$error_2 = 0.100°C$$

$$error_2\% = 0.459\%$$

The percent error signifies that the measurements recorded by the second student are more accurate.

Recording data: Keeping a good notebook

Even if you carry out a perfect experiment, unless you keep a good record of your data and your observations while you are doing the experiment, you will not be able to write a good report. This thorough and accurate laboratory notebook-maintaining skill will also help you later in life, working as a scientist or an engineer.

You need to keep a separate notebook to attach the material safety data sheet (MSDS) of the chemicals used, write the purpose of the experiment, your observations, and data. The guidelines below could assist you to maintain a laboratory notebook:

- Make sure you have the MSDS for all the chemicals that will be used during the experiment and attach them to your laboratory notebook in the section allocated for the relevant experiment.
- You should ensure a legible handwriting, as sometimes even you may not be able to read your own handwriting.
- Write all your observations. Things that may seem very trivial and easy to remember may not be remembered at the time of report

writing. They may contain diagrams or drawings. Even taking pictures and video recording would be helpful to remember what had happened during the experiment.

- If you have any difficulty during the experiment, you should write them down. This information can then be used to explain some discrepancies that may exist between the expected values and your experimental results. Also, if you have a record of problems that arose and how you were able to resolve it, you can use this knowledge elsewhere when a different problem arises.
- Write down the details of the equipment you are using and any systematic error points you suspect, which may then be used in your error discussion while writing the laboratory report.

Seek help from your assistants and instructors if you have any doubts.

Presenting data: Writing a laboratory report

After each experiment, generally a report is submitted, which provides information about the background of the experiment, materials, experimental procedures and conditions, recorded data, observations, calculations, if any, results, and a discussion. These items could be considered as a guideline or a template to prepare a laboratory report. The report is then generally submitted to teaching assistants or to the instructor, at most 1 week after the experiment was conducted. Generally, a report should consist of the following details, presented in the following order.

Cover page

It shows the course code, the course title, the title of the experiment, the name of the student performing the experiment, partner's name, if any, the date on which the experiment was performed and the date of submission of the report, and the authority to whom the report is submitted to.

Table of contents

This section should contain the contents of the report with page numbers. This enables the reader to follow the report and understand the structure of the work.

Theoretical background

This part should be an independent research of the topic. This information should be gathered from textbooks, Internet, etc., which should be referenced accordingly. The theory tested should be explained in detail, along with relevant equations used for calculations. Reporting previous published works on the topic might be helpful to show what is accomplished with this work. Also in this section, a clear representation of the purpose of the experiment should be given. Any information relevant to experimental set-up could also be presented in this section.

Materials

Here, all the equipment that have been used, including the chemicals, should be made known. If any specific concentration or storage conditions exist, this information should also be reported for future reference.

Procedure

In this section, the procedure should be written down step by step. It is important to write what you have done. Providing sufficient detail about the procedure is crucial to repeat the experiment. The experimental conditions, that is, temperature, settings, and prior calibrations should be presented. Generally, passive voice is preferred instead of the active voice.

Experimental data, observations, and calculations

During the experiment, any kind of observation that is to be considered important should be recorded on the experiment's data page. Observations section should include these records.

In some institutions, a laboratory notebook is required to record the data collected during the experiment. The carbon copy laboratory notebooks allow creating duplicated sheets; one copy can be returned to the instructor. For other composition-type laboratory books, the data should be written in ink during the experiment and should be signed on that date by an assistant. Photocopy the data sheet from your laboratory notebook and attach to your laboratory report. Mistakes committed during the recording of data are common, so if you make a mistake just cross out the wrong value and write the correct value above/next to the wrong value.

Calculations should follow the observation and data. Each step should be shown clearly so that it can be followed easily. If calculations are to be performed on a set of data, it is sufficient to show the calculation for one datum and indicate the remaining results in a tabular form. The proper units should accompany all numerical values.

Graphs

Graphs should be drawn using appropriate software, for example, Excel, Origin, Kaleidagraph. The title of the plot and labels of the axes with appropriate units should be shown in the graph. If you use a fitting to your data, you need to show the corresponding equations.

Results and discussion

This section is not a summary of the experiment. It consists of the following parts:

- The results obtained in calculations section should be presented in a tabular form for comparison of different cases.
- A statement indicating whether the observations are in accordance with the predictions of the theory.
- An error discussion for the results: A list of possible sources of random and systematic errors.
- A clear statement of the best numerical estimate and uncertainty in proper units, rounded to the correct number of significant figures.
- A comparison of the result with an accepted reference value should be given. Examples to references from a book, journal, or website are given below. If no literature value can be found, the list of all relevant sources searched should be supplied. If Internet is used to obtain the information, ensure that the source is reliable and credible.

References

Reference managing software are suggested to deal with more than a few cited publications. Use of such software will help one to organize the references in the document, as well as to follow some rules to ensure that the style is consistent. Below is a summarized list of American Chemical Society reference style.

Books and encyclopedias

Hiemenz, P. C. and Rajagopalan, R., *Principles of Colloid and Surface Chemistry*. 3rd Edition; Marcel Dekker: New York, 1997: pp 31–45.
Klempner, D., Sendijarevic, V., and Aseeva, R., *Handbook of Polymeric Foams and Foam Technology*. Hanser Gardner Publications: Kempten, 2004: pp 52–55.

Books section

Schunk, P. and Scriven, L., Surfactant Effects in Coating Processes. In *Liquid Film Coating: Scientific Principles and Their Technological Implications*, Springer: The Netherlands, 1997: pp 495–536.

Chapter in an edited book

Gleditsch, K. and Ward, M., Visualization in International Relations. In *New Directions for International Relations: Confronting the Method-of-Analysis Problem*, Mintz, A. and Russett, B.M., Eds. Lexington Books: New York, NY, 2005: pp 65–91.

Dissertations and theses

Zeng, C. Polymer Nanocomposites: Synthesis, Structure and Processing. Ph.D. Thesis, Ohio State University, 2003.

Journal articles

Bucak, S., Cenker, C., Nasir, I., Olsson, U., and Zackrisson, M., Peptide Nematic Phase. *Langmuir* 25, 2009: 4262–4265.

Fornara, A., Johansson, P., Petersson, K., Gustafsson, S., Qin, J., Olsson, E., Ilver, D., Krozer, A., Muhammed, M., and Johansson, C., Tailored Magnetic Nanoparticles for Direct and Sensitive Detection of Biomolecules in Biological Samples. *Nano Lett.* 8(10), 2008: 3423–3428.

Conference proceedings

Ozisik, R., Yang, K., and Liu, T. In *Novel Nanostructures Created by Supercritical Fluid Processing of Polymers*, American Physical Society March Meeting, Denver, USA, March 5–9, 2007.

Baysal, N., Unsal, B., and Ozisik, R. In *Interaction of Surface Modified Carbon Nanotubes with Supercritical Carbon Dioxide*, American Physical Society March Meeting, Baltimore, USA, March 13–17, 2006.

Material safety data sheets

Acetone; MSDS No. 9927062 [Online]; ScienceLab: Houston, TX, http://www.sciencelab.com/msds.php?msdsId=9927062 (accessed January 5, 2013).

Websites

Shenogin, S. and Ozisik, R. XenoView v.3.7.8. http://xenoview.mat.rpi.edu/ (accessed December 20, 2012).

References

Bauer, H.H., *Scientific Literacy and the Myth of the Scientific Method*. University of Illinois Press: Urbana, 1992.

Greenfield, T., *Research Methods*. Wiley: New York, 1996.

Hamming, R.V., The Unreasonable Effectiveness of Mathematics. *Am. Math. Mon.* 87(2), 1980: 81–90.

Lenk, C., Ethics and Law of Intellectual Property: Current Problems in Politics, Science and Technology. In *Applied Legal Phiosophy Series*, Lenk, C., Hoppe, N., and Andorno, R.L., Eds. Ashgate Publishing: Hampshire, 2007.

Shrader-Frechette, K.S., *Ethics of Scientific Research*. Issues in Academic Ethics Series. Rowman & Littlefield: Lanham, 1994.

Smith, R.V., *Graduate Research: A Guide for Students in the Sciences*. University of Washington Press: Seattle, 1998.

Wilson, E.B., *An Introduction to Scientific Research*. McGraw-Hill: New York, 1952.

chapter two

Characterization techniques

Surface tension measurements

Seyda Bucak

Introduction

The surface tension of a liquid is responsible for everyday phenomena such as the formation of spherical drops of rainwater and the blowing of soap bubbles. It is responsible for the adhesion of liquids onto other surfaces and also for the contraction of liquids on surfaces in forming spherical drops.

The surface energy of a liquid can be determined by measuring the surface tension of the liquid as a function of temperature. Measuring the change in surface tension upon addition of a surface-active agent allows for the determination of the effective cross-sectional area of the surface-active molecule and in the case of surfactant assembly, it allows for the determination of the critical micelle concentration.

Surface tension can be defined as the amount of work needed to create (expand) a surface per unit area. It arises from the fact that molecules in the bulk and on the surface have different environments, whereas the molecules on the surface are not only in contact with each other, but also with molecules at the interface of the other phase. This can be visualized as if the molecules at the liquid surface form a liquid film above the bulk liquid (Figure 2.1). This creates a certain "discomfort" for the molecules at the interface, which is quantified as surface tension.

When the molecules are in the bulk, they experience several intermolecular forces with their neighbors. However, the net force on the molecules almost always amounts to zero. On the other hand, when a molecule is on the surface, as it interacts with air or with molecules of another kind, the net force is not zero, resulting in "tension." Therefore, it requires energy (i.e., work) to bring a molecule from the bulk to the surface (or interface).

Creating a new surface does require work, the amount of work being proportional to the area created, which can be expressed as

$$w = \gamma \Delta A \qquad (2.1)$$

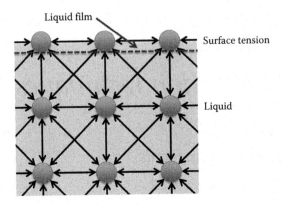

Liquid film

Surface tension

Liquid

Figure 2.1 The schematic representation of surface tension.

where γ is the proportionality constant, the so-called surface tension. Here, surface tension is represented as work/length².

On the other hand, the same phenomenon can also be visualized slightly differently. Let us say that there is a film formed by dipping a wire loop into a liquid (Figure 2.2). Films have very high surface areas compared to their volume and therefore have a tendency to spontaneously reduce their surface area.

If the loop has a wire that can slide as shown in Figure 2.2, then the wire moves in the direction of reducing the film surface. In order to prevent this, a force has to be applied. The amount of opposing force required is proportional to the surface tension of the liquid as shown below:

$$\gamma = \frac{F}{2l} \qquad (2.2)$$

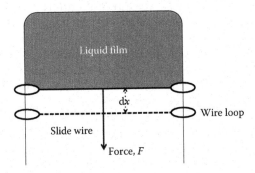

Liquid film

dx

Wire loop

Slide wire

Force, F

Figure 2.2 Liquid film formation by dipping a wire loop.

where *l* is the length of the film and 2 a factor due to the fact that a film has two sides. Here, the surface tension is expressed in terms of force/length.

These two approaches can be merged as follows. The amount of work required can be expressed as

$$dw = F\, dx \tag{2.3}$$

where $F = 2l\gamma$ resulting in $dw = \gamma 2l\, dx$, which can be resumed to

$$w = \gamma \Delta A \tag{2.4}$$

form as stated in Equation 2.1.

At constant temperature and pressure, non-expansion work is equal to Gibbs free energy, and surface tension is expressed as a thermodynamic quantity:

$$\gamma = \left(\frac{dG}{dA} \right)_{T,P} \tag{2.5}$$

where surface tension is the increase in Gibbs free energy per unit increase in the area.

Surface tension measurements, in general, can be classified as static and dynamic. For pure liquids, static surface tension measurements are evident. However, when there is a surface-active additive (surfactant) present in the medium, surface tension becomes a dynamic process as surface gets populated over time, the surface tension changes. Once the surface gets populated and reaches equilibrium, surface tension values level off. For each additive, equilibration time is different. The change in the surface tension as this equilibrium is reached is monitored by dynamic surface tension. The dynamic process of surfactant adsorption is very fast for common surfactants at most concentrations of interest, therefore time-resolved static measurements cannot be performed and special instrumentation for dynamic surface tension is required. Many interfacial phenomena that involve bubbles, menisci, thin films, and drops are affected by the equilibrium of dynamic surface tension, thereby making these measurements very important.

Static surface tension measurements

There are several methods to measure the static surface tension. The choice of method depends on the nature of the interface, the rheology of the liquids, the range of surface tensions to be measured, temperature and pressures of interest, speed, accuracy, precision, cost, etc.

Shape and pressure methods

Most surface tension measurement methods use the fact that surface tension keeps a liquid in the shape of a drop to reduce its surface area. Inside a drop of liquid, the pressure on the concave surface from the inside is higher than that on the outside of the convex surface.

The difference between these two pressures depends on the interfacial tension and the surface curvature of radii. The shape of a drop is determined by its radii of curvature R_1 and R_2 (Figure 2.3). The relationship between the interfacial pressure and these radii of curvature is described by the Young–Laplace equation:

$$\Delta P = \gamma \left(\frac{1}{R_1} + \frac{1}{R_2} \right) \tag{2.6}$$

where ΔP is the interfacial pressure difference, γ is the interfacial tension, and R_1 and R_2 are the radii of curvature of the surface.

For a sphere, $R_1 = R_2 = R$ and the equation reduces to

$$\Delta P = \frac{2\gamma}{R} \tag{2.7}$$

For a cylinder $R_1 = R$ and $R_2 = \infty$ and the equation becomes

$$\Delta P = \frac{\gamma}{R} \tag{2.8}$$

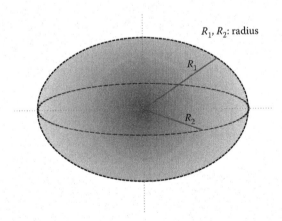

Figure 2.3 The schematic representation of a liquid drop described by the Young–Laplace equation.

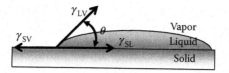

Figure 2.4 The curvature of a liquid drop formed on a surface.

Drop shape analysis

In this method, a drop of liquid is formed and the shape of the drop is recorded by a camera. The recorded image is then analyzed by fitting the shape of the drop to the Young–Laplace equation (Figure 2.4), which relates interfacial tension to drop shape to reveal the surface tension (Woodward, 2008). Often, this calculation is automatically done by the instrument's software.

This is an easy method that requires very little amount of liquid. In the calculations, the drop is assumed to be symmetric about the central vertical axis. Measurements can be performed on a sessile drop, where the drop lies on a surface, or on a pendant drop, where the drop is suspended in air, often at the tip of a needle (Figure 2.5).

During the measurements, the only forces shaping the drop are surface tension and gravity. The Young–Laplace equation is based on two radii of curvature, thus on distortion of the drop. When a pendant drop is used, the distortion of the drop is due to gravity. In order for this distortion to take place, there has to be a pressure difference between the top and the bottom of the drop, which requires that the drop has a certain height. Depending on the surface tension of the liquid that is being measured, the drop size should be adjusted to create this pressure difference sufficient to distort the drop. To image a drop, it should remain attached to the tip of a needle. This is achieved by slowly pushing the fluid out. When more viscous liquids are used, the pumping of the fluid should be slowed down further.

To study the liquid–liquid interface, drop-down or bubble-up (inverted) method can be used. To use the bubble-up method, the fluid must be transparent. For the liquid–gas interface, drop-down method should be used.

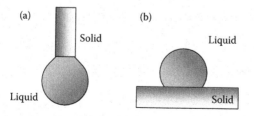

Figure 2.5 Presentation of (a) pendant and (b) sessile drop formation.

If the surface tension should be measured at a temperature other than ambient temperature, then the chamber where the drop is should be heated to the desired temperature, after which the image is recorded.

If a fluid with very low viscosity, such as molten glass, is to be measured, then special heating devices should be incorporated into the existing measurement system.

Drop weight (stalagmometer)

The measurement principle of a stalagmometer uses the relationship between the weight of a drop of liquid and its surface tension. Stalagmometer is a vertically placed glass tube with a wide liquid compartment in its middle section. The end of the tube, where the drop is released, is narrow (Figure 2.6).

When a drop forms at the end of the tube, it gets released when its weight equals the surface force at the end of the tube:

$$mg = 2\pi r \gamma \qquad (2.9)$$

where m is the mass, g the gravitational acceleration, r the radius of the stalagmometer, and γ the surface tension. However, in practice, for the same stalagmometer used, g and r being constant, a more useful relationship comes from the fact that the ratio of the mass and surface tension is constant for all liquids:

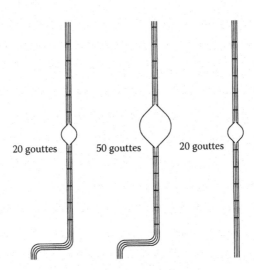

20 gouttes   50 gouttes   20 gouttes

Figure 2.6 Stalagmometers may have different middle volumes or different end configurations. The first two are more suitable for liquids with high viscosity, whereas the third is used for low-viscosity fluids.

$$\frac{m_1}{\gamma_1} = \frac{m_2}{\gamma_2} \tag{2.10}$$

By using a liquid with known surface tension, the surface tension of the unknown liquid can be calculated from the ratio of the masses of the two liquids. In order for this method to give accurate results, a sensitive balance should be used. Measuring the weight of one drop may be too difficult, so 10–20 drops may be measured to increase the precision.

Maximum bubble pressure

This is a simple, yet effective way to measure surface tension. Surface tension is determined from the value of the maximum pressure needed to push a bubble out of a capillary into a liquid, against the Young–Laplace pressure difference, ΔP. In this method, a bubble is formed at the end of a capillary with a known radius inside the fluid, the surface tension of which is to be measured, and the pressure required to create the bubble is measured (Erbil, 2006). As the bubble forms and expands, it creates a new surface for surfactants to adsorb (Figure 2.7). This method is most useful when surfactant adsorption is fast, which is often attained around critical micelle concentration values of surfactants.

Pressure is recorded during the bubble formation and detachment. Pressure becomes maximum when the bubble radius is the same as the capillary radius. Beyond that, bubble becomes unstable and detaches itself from the capillary.

The surface tension is calculated according to the Young–Laplace equation

$$\gamma = \frac{(P_{max} - P_0)r}{2} \tag{2.11}$$

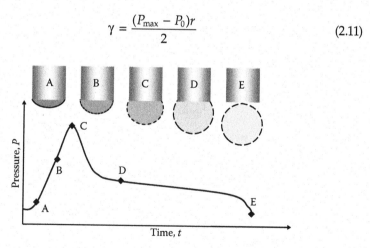

Figure 2.7 Pressure change with respect to drop size.

where P_0 is the hydrostatic pressure in the capillary that is because of it being immersed in a liquid, P_{max} the maximum pressure, and r the inner radius of the capillary.

Spinning drop
This method is particularly suitable for measuring liquid–liquid interfacial tension or to measure interfacial tensions of highly viscous liquids with precise temperature control. It is very good for measuring the interfacial tension of liquids of very low interfacial tension, such as microemulsions and miniemulsions.

A liquid is placed in a horizontal tube that can be spun about its longitudinal axis. Another liquid drop (less dense) is suspended in this liquid. During the rotation, the drop will reach the center of the tube and will get elongated as shown in Figure 2.8.

As the speed of rotation increases, the drop further elongates. At high rotational speeds, the drop shape becomes cylindrical as centripetal force overcomes the surface tension forces that keep the drop spherical. The equation of correlated surface tension with angular velocity is given by the Vonnegut equation:

$$\gamma_{1-2} = \frac{1}{4} r_0^3 \Delta\rho\omega^2 \tag{2.12}$$

where ω is the angular velocity of the spinning horizontal tube, $\Delta\rho$ the density difference between the drop and the surrounding liquid, and r_0 the radius of the drop. This equation is valid when the drop is in equilibrium and its length is at least four times longer than its diameter. The images of the drop are taken by a camera.

Alternative methods such as noncontact techniques for measuring in situ surface tension of liquids using ultrasonic waves have also been developed (Cinbis and Khuri-Yakub, 1992).

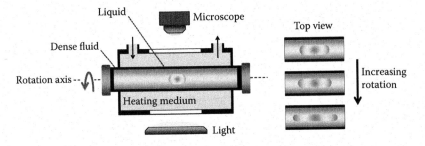

Figure 2.8 The schematic representation of spinning drop method: as the rotation speed increases (centrifugal force increases), the liquid elongates.

Capillary action

Capillary action is the result of adhesion of liquid to the walls of its container and surface tension. Adhesion causes an upward force on the liquid at the edges, while the surface tension holds the surface together (Figure 2.9). As a result of surface tension, instead of only the edges of the liquid moving upwards, the whole liquid surface moves up (Cengel and Cimbala, 2009).

Capillary action occurs when the adhesion is stronger than the cohesive forces between the liquid molecules. The extent of height to which the liquid will rise in the tube is determined by surface tension.

The capillary force is given by

$$F_{capillary} = \gamma 2\pi R \cos \theta \qquad (2.13)$$

where γ is the surface tension and R the radius of the tube.

The height h to which capillary action will lift the liquid depends on the weight of the liquid:

$$weight = mg = \rho Vg = \rho g(\pi R^2 h) \qquad (2.14)$$

Upon equating these forces, the force balance is given as

$$\gamma 2\pi R \cos \theta = \Delta \rho g h \pi R^2 \qquad (2.15)$$

where $\Delta\rho$ is the density difference between the liquid and the vapor (where ρ of the vapor can be neglected) and g the gravitational acceleration. If interfacial tension is to be measured, then the density difference between the two liquids should be considered. By measuring the height of the liquid in a capillary, surface tension can be calculated from the following equation:

$$\gamma = \frac{\Delta \rho g h R}{2 \cos \theta} \qquad (2.16)$$

When complete wetting occurs, we have that $\cos \theta = 1$.

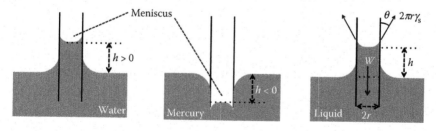

Figure 2.9 The capillary rise in water and capillary fall in mercury.

The same equation can be obtained by equating the Laplace pressure across the interface for a spherical drop:

$$\Delta P = \frac{2\gamma \cos\theta}{R} \qquad (2.17)$$

to hydrostatic pressure:

$$\Delta P = \Delta\rho gh \qquad (2.18)$$

Equation 2.16 can be derived from this equality also.

This is a simple and well-understood method for measuring surface tension. Often glass capillaries are used as they are completely wettable (contact angle of the liquid on the capillary wall is 0, resulting in a complete hemispherical meniscus) by most liquids due to its high surface free energy. However, special attention must be paid while cleaning the capillary tubes as even a small amount of impurity would introduce a high error in the measurements. Often this method is used for the surface tension measurements of pure liquids. The capillary tubes must be completely circular and even across the whole tube for accurate measurements. Often, the height of the liquid column in a capillary tube is measured against a reference liquid to reduce errors in the determination of tube radius.

Force methods

When a solid object is in contact with a surface such as air–liquid or liquid–liquid, a meniscus is formed. Force methods involve the measurement of the weight of the meniscus.

Du Nuoy ring

In this method, the ring is attached to a handle as shown in Figure 2.10. Often the solution is raised toward the ring until the ring is just in contact with the surface. Once this contact is established, the solution is lowered slowly, stretching the liquid film underneath the ring.

There is a maximum amount of force (F_{max}) that can pull the film before it breaks (Figure 2.11).

Once this maximum force is measured, the net force on the film is calculated by subtracting the weight of the volume of liquid lifted beneath the ring, F_V, from F_{max}. Therefore, surface or interfacial tension can be calculated using:

$$\gamma = \frac{F_{max} - F_V}{L \cos\theta} \qquad (2.19)$$

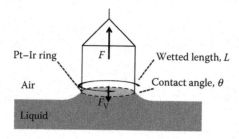

Figure 2.10 The schematic representation of the ring method.

where γ is the surface or interfacial tension, F_{max} the maximum force, F_V the weight of the volume of liquid lifted, L the wetted length, and θ the contact angle. Because the contact angle θ decreases as the ring is pulled away from the surface and becomes $0°$ at F_{max}, the value of cos θ becomes 1.

However, during this measurement, the surface is disturbed, resulting in difficulty in measuring equilibrium surface tension. The rate of pull of the ring out of the surface and the rate of equilibration of the surface play a role in determining F_{max}. Usually, slowing down the pull results in surface tension values to approach more accurate results. For pure liquids, this does not present a problem: however, when surfactant molecules are present, the disturbance of the surfactant molecules at the surface results in different orientation and concentration of surfactant molecules at the surface, resulting in inaccurate measurement. This makes the ring method not so suitable for the measurement of surface tension of surfactant solutions.

Rings are often made of platinum–iridium alloys to have very high surface tensions. They are very delicate and it is difficult to keep the rings free

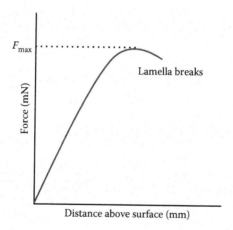

Figure 2.11 Change of force with ring distance.

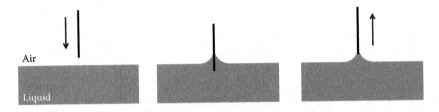

Figure 2.12 The schematic representation of Du Nuoy–Padday rod.

of bending. Even a slight bending causes the surface to be pulled unevenly because it will not be perfectly parallel to the surface for all contact areas.

Because the curve of the film is greater at the inside of the ring than at the outside, the measured maximum force does not agree exactly with the actual value. Different correction methods are applied based on the measurement with different ring diameters. This correction is often done automatically by the instrument.

Du Nuoy-Padday rod
In this method, instead of the ring, a small rod is used. It is immersed in the solution and the thin rod is gently pulled out, creating a meniscus as shown in Figure 2.12.

Because there is no ring, the problem of the difference of inner and outer ring is eliminated. Its small size allows the use of solutions with small volumes. This rod is often made of composite materials, hence are not easily bent like the ring. This method can only be used to measure surface tension and not interfacial tension.

Wilhelmy plate
This method is very similar to the abovementioned ring method, except that instead of the ring, a flat piece of wire, often made of platinum, is used to form a meniscus. Across the perimeter of the plate, the meniscus is formed as shown in Figure 2.13. As the plate is placed right at the surface

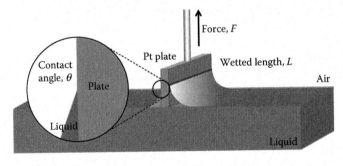

Figure 2.13 Wilhelmy plate method.

of the liquid, meniscus is formed without the need to pull the plate above the surface. This eliminates the aforementioned problem while using the ring method.

The plate does not have to be pulled above the surface to form the meniscus. The plate can rather be placed right at the surface of the liquid being measured and is not moved while surface tension is being measured, eliminating the surface equilibration problem that occurs in the case of Du Nuoy-Padday ring. The plate must be completely wetted before the measurement to ensure that the contact angle between the plate and the liquid is zero. It can be allowed for sufficient time (usually about a minute) for surface equilibration of surfactant solutions and surface tension can be deduced from the net force.

Since, the Wilhelmy plate is a block of platinum, it does not bend as easily as the ring, therefore, using the plate instead of the ring makes the measurement of surface tension much easier and more accurate. However, if interfacial tension (between two liquid surfaces) is to be measured, ring method presents some advantages. It is more difficult to keep the plate flush with the interface. Moreover, a meniscus is formed only by pulling the interface. This brings in the problem of interface not being in equilibrium even with the plate. On the other hand, using the ring, the wetted length is much longer than the plate, increasing the measured force, making the measurements more accurate.

Dynamic surface tension measurements

In a surfactant solution, when a new surface is created, it takes time for the surface to reach the equilibrium surface tension before the amount of surfactant at this surface equilibrates. This transport of surfactants to the fresh surface can take place within milliseconds or can take days, depending on the surfactant type (Ravera et al., 1993; Liggieri et al., 1996; Eastoe and Dalton 2000; Erbil, 2006).

Measuring dynamic surface tension is important for understanding emulsification, foaming, surface rheology, coating flows, etc. These processes are present in many applications ranging from technological to medical (Notter and Finkelstein, 1984; Valentini et al., 1991; Miller et al., 1994; Schunk and Scriven, 1997). From the point of view of surface science, the rate of surfactant adsorption or surface excess concentration with time can be determined from these measurements.

Measurements of dynamic surface tension lead to the mechanisms governing the adsorption behavior. Two main mechanisms play roles: one approach is that the rate of surfactant adsorption at the interface is determined by diffusion of monomers in the bulk to the interface. The adsorption at the interface is thought to be very fast, given that the monomer is sufficiently close to the surface. This is the diffusion-controlled mechanism. However, the other mechanism assumes the rate-determining step to be the adsorption of the monomer at the interface rather than the transport of the monomer in the bulk to near the surface. Often what is observed is the mixed model, where as the concentration of surfactant already adsorbed at the interface increases, an adsorption barrier becomes more pronounced and overrides the diffusion-controlled mechanism. The aforementioned adsorption barrier may be due to a potential energy barrier, orientation, or statistically the surfactant finding an empty place to adsorb.

Most widely used techniques to measure dynamic surface tension are the maximum bubble and spinning drop methods.

Maximum bubble pressure
The principles of measuring surface tension using maximum bubble pressure were discussed in the previous section. To measure dynamic surface tension using this method, a gas bubble is continuously formed at a steady rate and released at the end of the capillary. Precisely measuring the pressure using a pressure transducer inside the bubble (or drop) throughout the formation and detachment process allows for the evaluation of dynamic surface tension over a range of growth rate of a bubble.

Spinning drop
The principles of measuring surface tension using spinning drop were discussed in the previous section. As the images of the drop are taken by a camera, measurement of drop diameter with respect to time allows for the determination of dynamic interfacial tension.

Other methods

There are also other methods to measure dynamic surface tension such as oscillating jet, inclined plate flow cell, falling meniscus, growing drop or drop weight, and dynamic capillary method. Owing to their limitations, these methods are not widely used and most commercial tensiometers work with maximum bubble pressure principle.

References

Cengel, Y.A. and J.M. Cimbala. *Fluid Mechanics: Fundamentals and Applications*. 2nd Ed. McGraw-Hill Higher Education, 2009.
Cinbis, C. and B.T. Khuri-Yakub. A noncontacting technique for measuring surface tension of liquids. *Review of Scientific Instruments* 63, 1992: 2048–50.
Eastoe, J. and J.S. Dalton. Dynamic surface tension and adsorption mechanisms of surfactants at the air–water interface. *Advances in Colloid and Interface Science* 85(2–3), 2000: 103–44.

Erbil, H.Y. *Surface Chemistry of Solid and Liquid Interfaces.* Delhi, India: Blackwell Publishing, 2006.

Liggieri, L., F. Ravera, and A. Passerone. A diffusion-based approach to mixed adsorption kinetics. *Colloids and Surfaces A: Physicochemical and Engineering Aspects* 114, 1996: 351–59.

Miller, R., P. Joos, and V.B. Fainerman. Dynamic surface and interfacial tensions of surfactant and polymer solutions. *Advances in Colloid and Interface Science* 49, 1994: 249–302.

Notter, R.H. and J.N. Finkelstein. Pulmonary surfactant: An interdisciplinary approach. *Journal of Applied Physiology* 57(6), 1984: 1613–24.

Ravera, F., L. Liggieri, and A. Steinchen. Sorption kinetics considered as a renormalized diffusion process. *Journal of Colloid and Interface Science* 156(1), 1993: 109–16.

Schunk, P.R. and L.E. Scriven. Surfactant effects in coating processes. In Kistler, S.F. and P.M. Schweizer, (eds.) *Liquid Film Coating: Scientific Principles and Their Technological Implications.* 495–536: Springer: Netherlands, 1997.

Valentini, J.E., W.R. Thomas, P. Sevenhuysen, T.S. Jiang, H.O. Lee, Y. Liu, and S.C. Yen. Role of dynamic surface tension in slide coating. *Industrial & Engineering Chemistry Research* 30(3) 1991: 453–61.

Woodward, R.P. Measurements using the drop shape method (http://www.Firsttenangstroms.com). 2008.

Viscosity/rheological measurements

Patrick Underhill

Introduction

Fluids versus solids and viscoelasticity

When applying a force onto a fluid (either liquid or gas), the material will continually deform. This deformation requires that parts of the fluid are either sliding past one another or stretching apart. The viscosity of the fluid is a measure of the fluid's resistance to this deformation just as friction resists the relative motion of two solids. When applying a force to an elastic solid (such as a rubber band), the material may deform to some extent, but then may stop deforming. The solid has the properties of the past state in which there was neither force nor deformation, and would, by all means, return to that state. Materials that possess the characteristics of both elastic solids and viscous fluids are called viscoelastic.

Viscoelastic fluids typically have fading sustenance; they can sustain only for a particular period of time λ called the relaxation time, but because of stretching they lose the property after that said period of time. If a material experiences the force or is under an observation time t_{obs}, then the material will certainly behave as an elastic solid if $t_{obs} < \lambda$ and as a viscous fluid if $t_{obs} > \lambda$. This ratio is called the Deborah number De = λ/t_{obs}. For many fluids made of small molecules, the relaxation time is very small, leading to a small De, and remains a viscous fluid under most situations. In polymeric systems or colloidal suspensions, the larger sizes of the constituents lead to larger relaxation times (longer memory) and can lead to many fluids that are being viscoelastic.

Material functions

In an incompressible Newtonian fluid, the resistance of the fluid to deformation is quantified by a single parameter called the viscosity μ. This parameter is dependent on the material, hence it is called a material property. Many polymeric systems or colloidal suspensions are non-Newtonian, so are quantified by more than just a single material property. The resistance to deformation can depend on the material's history of deformation or the rate at which the deformation is taking place (e.g., the viscosity can depend on the shear rate). The function, which quantifies how this resistance varies, is material-dependent, hence is a material function.

Kinematics

The study of how materials flow is called Rheology. Rheology uses material functions to quantify how a material flows and also to compare them with other materials. Most material functions are defined from a kinematics viewpoint. Kinematics is the study of motion and the impact of the motion without any concern for what is needed to cause the motion. So, most of the material functions compare the stress in the material resulting from a prescribed deformation to a measure of the deformation. One important exception is creep measurements (not described in detail here) in which a prescribed stress is imposed and the deformation in response is measured and compared with the stress. Note that even for material functions based on kinematics, many rheometers are stress-controlled and use feedback control to impose the desired deformation.

Capillary flow

Flow through cylindrical capillaries is a common method for measuring the shear viscosity of fluids (Figure 2.14). For materials with a viscosity that does not depend on the shear rate, the velocity profile within the tube can be calculated exactly and related to the viscosity. By measuring both the forcing needed to cause the flow (e.g., using the pressure drop) and the deformation (e.g., using the flow rate), the viscosity can be calculated. Some viscometers use gravity to drive the flow and measure the time to fall a known distance instead of measuring a pressure drop and the flow rate, but the principles are the same. For fluids with a viscosity that depends on the shear rate, for which the velocity profile is not known

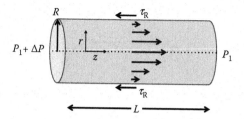

Figure 2.14 Illustration of flow through a capillary driven by pressure gradients (or gravity). These forces are balanced by shear stresses at the wall.

exactly, methods have been developed to extract the viscosity material function as a function of the shear rate.

Newtonian profile

For a pressure or gravity-driven flow through a cylindrical capillary of radius R, the laminar flow velocity profile for an incompressible Newtonian fluid is given by

$$v = 2U\left[1 - \left(\frac{r}{R}\right)^2\right] \qquad (2.20)$$

where $U = Q/\pi R^2$ is the mean velocity across the channel and Q is the volumetric flow rate. The mean velocity (or the flow rate) is related to the viscosity and pressure drop (or the gravitational force) in what is called the Hagen–Poiseuille equation:

$$\frac{\Delta P}{L} = \frac{8\mu U}{R^2} = \frac{8\mu Q}{\pi R^4} \qquad (2.21)$$

In the case of falling of a fluid through a capillary in which the flow is gravity-driven, the term $\Delta P/L$ is replaced by ρg, where ρ is the fluid density. If all other terms are measured, this equation can be used to calculate the viscosity.

Weissenberg–Rabinowitsch method

For a Newtonian fluid, the viscosity is independent of the shear rate in the capillary. If the viscosity depends on the shear rate, we need an alternative method to determine the viscosity and to determine the corresponding shear rate of the viscosity. The Weissenberg–Rabinowitsch method is a way to do this for an arbitrary incompressible fluid (Figure 2.15).

In a capillary flow, both the stress and shear rates vary from zero at the channel center to a maximum at the wall. The shear viscosity material

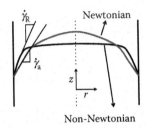

Figure 2.15 Illustration of the different flow profiles in a capillary for a Newtonian or non-Newtonian fluid. For a Newtonian fluid, the shear rate at the wall equals the apparent shear rate. For a non-Newtonian fluid, the true shear rate at the wall differs from the apparent shear rate.

function is defined as the stress divided by the shear rate. It is conventional to use the capillary flow to determine the viscosity of the fluid near the wall. This is because the stress at the wall ($r = R$) is given by $\tau_R = \Delta P(R)/2L$ or any fluid. The challenge is to determine the shear rate at the wall, since for a general fluid the velocity profile is unknown. The flow rate can be used to construct a shear rate, called the apparent shear rate, which would be the shear rate at the wall if the fluid were Newtonian $\dot{\gamma}_a = 4Q/\pi R^3$. For a Newtonian fluid, the viscosity is simply the ratio $\tau_R/\dot{\gamma}_a$. It is possible to calculate the exact shear rate at the wall $\dot{\gamma}_R$ by examining how $\dot{\gamma}_a$ changes when τ_R is changed, specifically from the slope on a log–log scale. By varying the pressure drop (wall shear stress) and flow rate (apparent shear rate), the viscosity as a function of the shear rate can be determined using

$$\eta(\dot{\gamma}_R) = \frac{4\tau_R}{\dot{\gamma}_a}\left(3 + \frac{d \ln \dot{\gamma}_a}{d \ln \tau_R}\right)^{-1} \qquad (2.22)$$

Drag flow

Another common method for causing deformation, instead of applying a pressure drop or body force from gravity, is to move one of the walls that will drag the fluid causing a flow. We will review the three most common geometries for which the drag flow is used to extract viscosity information.

Parallel plates

Consider a fluid between two parallel disks of radius R separated by a distance H (Figure 2.16). If the top disk (at $z = H$) is rotated at a constant angular velocity, Ω, while the bottom disk (at $z = 0$) is held stationary, the velocity of the fluid in the gap is $v = \Omega z r/H$. This will be true for any

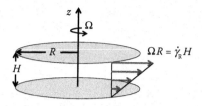

Figure 2.16 Illustration of the flow between two parallel plates. The shear rate at the edge of the plate is related to the angular velocity, radius, and separation of the plates.

fluid provided the gap is small enough and the additional considerations mentioned later are included. From this velocity profile, the shear rate is $\dot{\gamma} = \Omega r/H$ and varies from zero at the center to a maximum at the edge of the disk $\dot{\gamma}_R = \Omega R/H$.

Similarly, the stress will vary across the fluid. Instead of measuring this stress, it is conventional to measure the torque T, required to rotate the plate, which is related to the integral of the stress across the plate. Note the similarity with capillary flow for which the shear rate and stress varied throughout the fluid. A similar method can be used to extract the viscosity for a general fluid.

In a capillary flow, the stress at the wall was known exactly while the shear rate was calculated by examining a plot of apparent shear rate versus stress. In parallel-plate flow, it is conventional to calculate the viscosity of the fluid (which could be shear rate-dependent) at the shear rate at the edge of the disk $\dot{\gamma}_R = \Omega R/H$. This shear rate is known, but the stress is unknown and must be inferred by plotting the shear rate versus the torque on the top plate. In particular, the viscosity is obtained by

$$\eta(\dot{\gamma}_R) = \frac{T/2\pi R^3}{\dot{\gamma}_R}\left(3 + \frac{d\ln T/2\pi R^3}{d\ln \dot{\gamma}_R}\right) \tag{2.23}$$

Cone-and-plate

We saw that for the parallel-plate geometry the shear rate varied with distance from the center of the disk. This is because the linear velocity of the top plate increased with distance but the gap between the plates was constant. If the gap were also to increase with distance from the center, then the shear rate would have been constant. This is the case for the space between a flat disk and a cone (Figure 2.17). If the angle of the space between the plate and the cone is Θ and is small, then the shear rate will be constant and is equal to $\dot{\gamma} = \Omega/\Theta$. The torque on the cone T is related to the fluid stress (which is now constant since the shear rate is

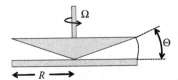

Figure 2.17 Illustration of the cone-and-plate geometry. The shear rate is constant throughout the fluid and is dependent on the angular velocity and angle of the cone.

constant). The ratio of the stress to shear rate gives the viscosity, which is given by

$$\eta = \frac{3\,T\,\Theta}{2\,\pi R^3 \Omega} \tag{2.24}$$

The cone-and-plate geometry has some key advantages over the parallel-plate geometry. The uniformity of the shear rate can be significant for materials that undergo large structural transitions as a function of shear rate. The uniformity also simplifies the calculation of viscosity, not requiring a plot to extract the log–log slope of torque versus shear rate. However, for high viscosity fluids, it can be more challenging to load the rheometer, compared to the parallel-plate geometry.

Couette flow

The final common geometry to produce shear flows is the Couette geometry. In this geometry, fluid fills the gap between two concentric cylinders. When the cylinders are rotated relative to one another, the fluid between them is sheared (Figure 2.18). If the gap between the cylinders is small, then the fluid velocity profile will be linear for any fluid even with the

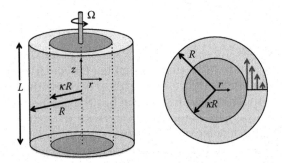

Figure 2.18 Illustration of the Couette geometry. The fluid is sheared by a relative angular velocity of the two concentric cylinders.

cylindrical geometry. The outer cylinder has a radius R and the inner cylinder has the radius κR, where κ is a parameter that is less than but very close to 1. As with the cone-and-plate geometry, the analysis is relatively simple since the shear rate remains constant throughout the fluid. The torque on the cylinder T is directly related to the stress in the fluid and the angular velocity Ω is directly related to the shear rate. Combining these quantities gives the viscosity as:

$$\eta = \frac{T(1 - \kappa)}{2\pi R^2 L \kappa^3 \Omega} \tag{2.25}$$

where L is the length of the inner cylinder, which arises from the surface area of the cylinder over which the stress is acting. This increased surface area is one of the key advantages of the Couette geometry for fluids with low viscosities. Fluid with a low viscosity may produce a very little torque that it falls below the sensitivity of the instrument. The larger surface area can increase the torque to fall within the range that can be accurately measured.

Additional considerations

The previous description has summarized some of the key methods for measuring the steady shear rheology of fluids. Some additional considerations should be borne in mind when applying the formulas and analysis. We discuss some of the most common here. More detailed descriptions can be found in the literature or texts devoted to the study of rheology.

Wall slip

When calculating the shear rate that a fluid experiences, the no-slip condition is typically assumed that the fluid next to a solid surface moves at the same velocity as the solid surface. This is not necessarily true in practice, especially for materials with structures at the colloidal scale. If the fluid slips across a surface, the shear rate and therefore viscosity will not be the same as calculated. In general, there are two ways of dealing with slip: altering the apparatus to eliminate slip or changing the analysis to account for slip. The most common method to eliminate slip is to use roughened plates with the scale of roughness comparable to the scale of microscale structure in the fluid. The roughness will prevent the microstructure from slipping past the plate. Alternatively, slip can be corrected by measuring the response for different geometries and comparing the measured viscosity. Typically, the gap between the plates, the cone angle, or capillary radius are changed in order to gauge the impact of slip.

Entrance effects

In a capillary flow, the formulas for the shear rate at the wall and the stress at the wall rely on the flow profile that is being the same over the whole length of the capillary. When the fluid first enters the capillary, there will be a different flow profile and therefore gives rise to different shear rates and stresses at the wall. These effects are typically analyzed in two ways: estimating their order of magnitude to justify ignoring them or varying the system to correct for them.

The entrance length for the fluid to reach its terminal velocity profile can be estimated for the laminar flow of Newtonian fluids. This entrance length is approximately $L_e \cong R(1 + 0.1\,Re)$, where R is the tube radius and the Reynolds number is $Re = 2UR/v$. If the length is much longer than the entrance length, then their effect on the viscosity calculations can safely be ignored. The entrance effects can be quantified by using capillaries with different lengths; the contributions from the steady regions will be proportional to the length, while the entrance effects will be independent of the length of the capillary.

Edge effects

In the parallel-plate and cone-and-plate geometries, the fluid at the edge of the plate and/or plates is open to the atmosphere. These edges can some-times influence the measured viscosities. In polymer melts, edge fracture can occur. In polymer solutions and colloidal suspensions, it is important to note that they might get absorbed at the interface and influence the measurements. When the fluid is being deformed, the fluid–air interface is also deformed and can resist that motion. In these cases, geometry such as a capillary flow is preferable since there are no fluid–air interfaces. Alternatively, surfactants can be used to displace the polymers or colloids at the interface, reducing the resistance to deformation of the interface.

Instabilities

In all of the methods described here, the fluid velocity profile is either known or inferred through other measurements. This velocity profile is the solution to nonlinear equations for which the nominal solution can be unstable, leading to a different flow profile. In a capillary flow, the most common instability is the transition to turbulence, which happens at a large enough Reynolds number. In the drag flows, the most common instability is a secondary flow that develops due to centrifugal forces. This will act to throw the rotating fluid out to the edge, with other fluid return-ing back to the center. This instability is not present in Couette flow when the outer cylinder is rotated since the faster fluid is already outside of the slower moving fluid. These instabilities will increase the stresses above those expected at the measured shear rates, causing the measured viscos-ity to be higher than the true value.

Torque sensitivity and instrument inertia

Viscosity and rheological measurements are accurate only within the restrictions of the apparatus performing the measurements. For example, in drag flows, if the torque is below the sensitivity of the rheometer to measure torques, the measurement will not be correct. This is particularly important in drag flows of fluids of low viscosity, which do not produce large torques (especially at low shear rates). If the torque drops below the level that can be accurately measured, the fluid can appear to be shear thinning (higher viscosity at lower shear rates) when it is simply a measurement error. In these cases, either the Couette geometry (with larger surface area) or capillary flow should be used to measure viscosity.

Another instrument limitation that can pose a threat is the inertia of the instrument. Oscillatory flows (not discussed here) can be used to characterize the viscous and elastic responses of a fluid. During oscillations, the inertia of the instrument will cause a time lag between the imposed stresses and the resulting deformations. It is important to include this effect so as to not attribute that lag to the material being measured. Inertia will not affect measurements in the steady flows described here.

Material segregation

The analysis provided here assumes that the material is homogeneous. In some polymeric or colloidal systems, the flow will cause the material to segregate into more concentrated and less concentrated regions. This will impact the measurements of stresses and viscosities.

Electrokinetic techniques

Marek Kosmulski

Introduction

This section is devoted to determination of electrokinetic potential. The electrokinetic potential cannot be measured directly, but it can be calculated from the quantities (e.g., electrophoretic mobility) that can be measured. The electrokinetic potential is an important quantity characterizing interfaces. The symbol ζ (zeta) indicates the electrokinetic potential.

Basically, the electrokinetic potential occurs at various types of interfaces including gas–liquid (e.g., air-bubbles dispersed in water) and liquid–liquid (emulsions), but it is mainly studied at the solid–aqueous solution interface. Theoretical and practical studies of the ζ potential at solid–aqueous solution interfaces outnumber the studies in other systems by a few orders of magnitude. The commercial equipment has especially been designed for solid–aqueous solution interfaces. Several manufacturers of zetameters claim that their equipment is also suitable for atypical systems, for example, dispersions in nonaqueous solvents or emulsions, but such measurements require state-of-the-art skills and knowledge. This section is focused on standard systems (solid–aqueous solution interfaces), although several principles outlined here may also be applicable to atypical systems.

Units, order of magnitude, accuracy, and precision

The ζ potential is expressed either in volts (V) or in millivolts (mV). The typical range is between –150 and +150 mV. For aqueous systems, ζ potentials substantially exceeding 150 mV in absolute value are very unusual and would probably be erroneous. The ζ potentials exceeding 150 mV in absolute value are often claimed for dispersions in nonaqueous solvents, which are atypical systems, and they are beyond the scope of this section.

Owing to the inherent nature of the interfaces (which are generally difficult to control), approximate character of models (equations) used in the calculations and limited accuracy of the numbers used in the calculations (particle size, ionic strength), the ζ potentials can hardly be determined with an accuracy of <1 mV. In practice, a reference (true) value of ζ potential is seldom known, so the number of decimal digits in the reported values of the ζ potential reflects precision rather than accuracy. Perhaps in a special case of an "easy-to-control" system, excellent equipment, and excellent operator, one decimal digit (e.g., 16.1 mV) may still be considered significant. The ζ potentials (in mV) reported with two (e.g., 16.12 mV) or more decimal digits are inherently incorrect, because two decimal digits suggest too high (overrated) accuracy.

Model of the interface

A model of an electrically charged solid–electrolyte solution interface is required to define the ζ potential. The ζ potential is observed in the electrokinetic phenomena, that is, when the solid moves with respect to the solution in tangential direction. The kinetic unit consists of the solid along with a thin (fraction of 1 nm) layer of solution, which contains solvent molecules and solutes (especially ions). The exact thickness of this layer (shear plane distance) may be an object of fascinating fundamental research, but there is no generally accepted approach to this problem, and we will define the ζ potential without referring to any specific value(s) of the shear plane distance. The electric potential in electrolyte solution as a function of the distance from the interface is plotted in Figure 2.19.

Figure 2.19 shows that the ζ potential is not the surface potential ψ_o, although these two potentials may be equal at certain circumstances. Both the ζ potential and the surface potential are defined against the electric

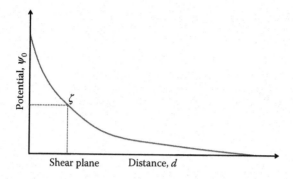

Figure 2.19 The electric potential in solution in the interfacial region.

potential in bulk solution (equal to zero by definition) as a natural reference. Typically, the ζ potentials are low in absolute value (up to 150 mV),
and surface potentials may be higher (several hundred mV) in absolute
value. Figure 2.19 presents one of the simplest models of a charged interface called diffuse layer model. Many other models have been developed,
but their discussion is beyond the scope of this section.

In order to observe electrokinetic phenomena (thus to determine the
ζ potential), the solid is required to be an insulator (show low electric conductivity) and the solution is required to show certain degree of ionic conductivity. Measurements in solutions of very low conductivity (including
very pure water) are considered as atypical and they may require spe-
cially designed equipment and skills. Perfect electric insulators do not
exist. From the viewpoint of electrokinetic measurements, it is important
that the conductivity of the solution exceeds the conductivity of the solid
by several orders of magnitude. The electrokinetic phenomena involving
conductive solids (in the sense explained above) are referred to as "electrokinetic phenomena of the second kind" (Mishchuk and Takhistov, 1995)
and they are beyond the scope of this section.

The conductivity of the solution in the range 10^{-3} to 0.1 S m^{-1} (corresponding to 10^{-4} to 10^{-2} M solution of 1–1 electrolyte) provides optimum
conditions for the determination of the ζ potential. This range is partially
because of intrinsic properties of colloidal systems and partially because
of the design of commercial zetameters. Many ζ potentials reported in
the literature have been determined for ionic conductivities beyond this
range. Possible difficulties related to high (low) conductivity of the electrolyte are discussed later in this section.

Basically, the ζ potential of relatively water-soluble solids can be determined, that is, its insolubility is not a pre requisite. Yet sparingly soluble
solids are easier to handle and most electrokinetic studies have been carried out for sparingly soluble materials, while measurements with water-
soluble solids are considered as atypical and difficult.

Sources of the electric potential in the interfacial region

The distribution of electric potential shown in Figure 2.19 is due to different affinities of ionic species (originally present in solution or resulting from dissolution of solid particles) to the surface. High affinity of cations to the surface leads to their adsorption (accumulation) at the surface and to positive surface charge and positive surface potential, as shown in Figure 2.19. High affinity of anions to the surface leads to negative surface charge and negative surface potential. The surface charge is neutralized by the countercharge in solution, which is equal to the surface charge in absolute value, but has an opposite charge. The surface charge is due to the charge of ions accumulated at the surface. The countercharge in solution is due to the excess of counterions, which results from depletion of solution from the co-ions. The surface charge is concentrated within a narrow range of distance from the interface (ordinate in Figure 2.19, distance equal to zero), while the countercharge has a "diffuse" character, and it is spread over a broader range of distances. The curvilinear distribution of electric potential shown in Figure 2.19 is due to mutual interaction of ions in the interfacial region. The counterions are attracted by the co-ions adsorbed at the surface, but they also repel each other, thus the layer of solution containing substantial excess of counterions is relatively thick. In Figure 2.19, the surface charge is not entirely neutralized by the counterions within the shear plane, that is, the kinetic unit, which consists of the solid and also the layer of solution next to it carries a net electric charge.

Figure 2.20 presents a distribution of the electric potential in solution for three different ionic strengths. Subscript 1 refers to the lowest and subscript 3 refers to the highest ionic strength. The surface potential and the distance of the shear plane are assumed to be constant (independent of the ionic strength). At a low ionic strength, the slope of the curve is low, and the ζ potential (ζ_1) is high in absolute value. At a moderate ionic strength, the slope of the curve is moderate, and the ζ potential (ζ_2) is moderate in

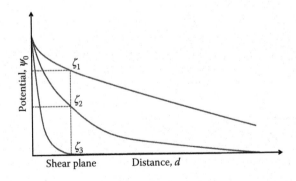

Figure 2.20 The ζ potential at various ionic strengths.

absolute value. At a high ionic strength, the slope of the curve is high, and the ζ potential (ζ_3) is low in absolute value. Figure 2.20 reflects a shift of the "center" of the countercharge toward the surface as the ionic strength increases.

Figure 2.20 indicates that, in principle, the ζ potentials are high in absolute value (±50 mV or more) at low ionic strengths (0.001 M of 1–1 electrolyte or less), and they are low in absolute value (±20 mV or less) at high ionic strengths (0.01 M of 1–1 electrolyte or more). Obviously, the ζ potential cannot be too high in absolute value when the surface potential is low in absolute value, even at a very low ionic strength. In contrast, a combination of a high absolute value of the surface potential and high ionic strength may result in a relatively high absolute value of the ζ potential.

Description of the system

The ζ potential is a property of the interface. It depends on the nature of the solid particles and on the composition of the solution. This property of the ζ potential is often overlooked, and the ζ potential is attributed to the solid only in some reports. For example, a statement that "the ζ potential of hematite particles equals to +50 mV" is inherently incorrect, because it refers to the particles only and it neglects the composition of the solution. A proper description should include information about the solution, for example, "the ζ potential of hematite particles in 0.001 M KNO_3 at pH 4 equals +50 mV."

In certain commercial dispersions (e.g., Ludox™), both the solid and solution are well defined (a short trade name includes full description of the system), and in such dispersions, detailed description of the solution composition is not required as long as the original dispersion is studied. However, any manipulation, which changes the composition of solution, for example, dilution with water or pH adjustment, may result in changes in the ζ potential in such commercial dispersions.

The adsorption properties of solids, that is, their affinities to particular solutes, depend on their external layer (shell, grafted compounds), while the nature of the bulk (core) is rather insignificant. Surface treatment can substantially change the value and even reverse the sign of the ζ potential. Many commercial materials have their trade names referring to the cores (which are the main components in terms of mass fraction, but have rather insignificant effect on the surface properties) rather than to the shells. This property is somewhat confusing. For example, a material composed of 96% of alumina (core) and 4% of silica (shell), called AluminaXYZ™, is likely to behave like silica rather than like alumina in terms of its ζ potential in contact with aqueous solutions.

The ζ potential is suitable for characterizing systems containing one type of solid, and, in principle, is not suitable for characterizing mixtures of different solids (naturally occurring or artificially prepared). Under certain special conditions, signals produced by particular types of solid particles (components of the system) can be distinguished and interpreted in terms of ζ potentials of particular components. However, the possibility of such a deconvolution is not granted, and interpretation of ζ potentials in systems involving mixtures of different solids is often based on speculative assumptions, and it requires special knowledge. An "overall" ζ potential of a mixture of different solids can be "technically" measured (and it is often reported in literature, e.g., for clays) but it does not have physical sense. Namely, the contribution of particular components to the overall signal depends not only on their mass fractions, but also on various properties of particles, for example, the size and shape of particles as well as the refractive index, which are difficult to control or interpret.

The composition of solution used to define the system is the actual (rather than initial) composition. The difference between the initial and the actual composition of the solution is often overlooked. "The ζ potential of hematite particles in 0.001 M KNO_3 at pH 4" means that the pH of the dispersion was measured after certain equilibration time (not necessarily at equilibrium) just before or after the electrokinetic measurement. Such a ζ potential is independent of the solid-to-liquid ratio, although various amounts of reagents may be necessary to adjust the dispersion pH to 4 at different solid-to-liquid ratios. In contrast, when the pH of a solution is pre-adjusted to 4, and then the particles are added, the final pH will be different from 4, and dependent on the solid-to-liquid ratio. Consequently, the ζ potential in a system defined by the initial composition of the solution depends on the solid-to-liquid ratio. Such ζ potentials in systems defined by the initial composition of the solution (determined before equilibration with the solid) are of limited significance, especially when the solid-to-liquid ratio is not reported.

The requirement that the actual concentrations of all components of the solution have to be included in the sample characterization is very demanding, especially in multi component solutions. Actually, only certain components substantially affect the ζ potentials of particles, while most other components do not. Minor components (present at low concentrations), which are not particularly surface-active, can usually be neglected in the description of the system.

Classical electrokinetic phenomena

Let us consider a kinetic unit composed of electrically charged solid and a thin layer of solution around it, which moves with respect to the rest of the solution in tangential direction. The relationship between the force and the flow (which can be of either mechanical or electrical) can determine the ζ potential. The phenomena in which electric force causes a

Table 2.1 Classical Electrokinetic Phenomena

		Electric force	Mechanical force
		Mechanical flow	Electric flow
Fixed solid	Moving solution	Electroosmosis	Streaming potential (current)
Fixed solution	Moving solid	Electrophoresis	Sedimentation potential (current)

mechanical flow or vice versa are referred to as electrokinetic phenomena. In the classical electrokinetic phenomena, the forces and flows (potentials, pressures) are constant over periods of time of >1 s. Their high-frequency counterparts (e.g., involving high-frequency alternated electric field) are called electroacoustic phenomena and will shortly be discussed later in this section.

Table 2.1 classifies the classical electrokinetic phenomena according to the applied force and observed flow. Two types of experimental systems are considered: "Fixed solid and moving solution" may refer to a capillary filled with a solution, in which applied electric potential results in flow of the solution (both in axial direction). The capillary is mounted in a horizontal position to avoid the interference from the flow caused by gravity. A capillary can be replaced by a channel formed by two parallel plates or by a bi continuous porous solid with pores filled with solution or by a closely packed plug consisting of many solid particles with solution in the interparticle space. Both porous solid and closely packed plug require a sample holder, which allows net flow of solution in one direction, and prevents net flow in any other direction. Porous solid and closely packed plug may be considered as equivalent to a bunch of parallel capillaries. "Fixed solution and moving solid" refers to solid particles dispersed in solution. The solution needs a vessel (sample holder).

Out of the four classical electrokinetic phenomena reported in Table 2.1, only one (sedimentation potential) is not suitable for the measurement of the ζ potential. In principle, in a dispersion of solid particles, which settles down under gravity, an electric current in vertical direction is induced, which can be measured and interpreted in terms of the ζ potential of particles. However, due to technical difficulties, sedimentation potential is of limited significance as an electrokinetic technique and will not be discussed here.

Electrophoresis

Electrophoresis is the most popular electrokinetic technique and there are many commercial instruments (zetameters) based upon electrophoresis available in the market. Modern zetameters are "black boxes," which accept a sample, and then produce a value of the ζ potential. The instrument software will adjust the conditions of the measurement to the character of the sample and even alert the user that certain samples may not be suitable for the ζ potential determination.

The experimental setup

The experimental setup used in electrophoresis is illustrated in Figure 2.21.

The gray color represents electrolyte solution in a container. The black round shape represents a kinetic unit, which consists of a colloidal particle with a thin layer of solution around it. The electrodes are arranged in such a way that the electric field lines are horizontal. This prevents the effect of sedimentation under gravity on the observed movement of a particle.

When the electric field is on, the particle moves in the direction of the oppositely charged electrode. The velocity of the particle (v) is proportional to the field strength (E). Thus, the electrophoretic mobility defined as

$$\mu = \frac{v}{E} \tag{2.26}$$

is independent of the field strength. Typical electrophoretic mobilities are on the order of 10^{-8} $m^2 V^{-1} s^{-1}$. In old-type zetameters, the velocities of several individual particles were measured manually, and the average velocity was taken as a representative number to estimate the average μ of all particles in the system. In a modern equipment, the measurement and averaging of the velocities of multiple particles is automatized. Individual particles in dispersion show different μ, although in systems well suited for electrophoretic measurement, the difference between the fastest and the slowest ones being not very significant. Yet, measurement of μ based on the velocity of a single particle is not advised. The movement of a

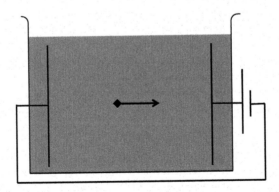

Figure 2.21 Electrophoresis.

particle under the Coulombic force overlaps with its Brownian motion. The latter is especially significant in very fine particles. Thus, the "electrophoretic mobility histogram" (a plot of the number of particles as a function of μ) is broad in fine particles and narrow in coarse particles. In instruments, which allow display of the "electrophoretic mobility histogram," its width may be used to estimate the particle size. There are factors other than Brownian motion, which affect μ of individual particles. For example, among particles having the same ζ potential, larger particles are usually faster than the smaller ones. In the so-called "monodispersed colloids" (powders and dispersions having a very narrow distribution of particle size), the effect of the particle size on μ of individual particles is rather insignificant, but in many commercial powders and dispersions, individual particles differ in size by several orders of magnitude, and this results in a substantial scatter in μ of individual particles. Finally, individual particles may differ in their surface properties, and in their surface potential, and they also have different ζ potentials.

The scatter in μ of individual particles in a typical sample results in limited precision of the values of ζ potential discussed in section "Units, Order of Magnitude, Accuracy, and Precision." It should also be emphasized that the collection of a representative sample from a lot of material (powder, dispersion) for ζ potential measurement may be a real challenge.

Systems suitable for electrophoretic measurements
It was silently assumed in Section "The Experimental Setup" that the particle moves in a horizontal direction under Coulombic force rather than in the vertical direction under gravity. This implies a stability of the dispersion against sedimentation. Unstable dispersions settle down (partially or totally) before the sample is loaded into a chamber, and a measurement can be carried out.

The velocity of sedimentation of primary particles depends on a combination of their size (fine particles are more stable) and specific density. Electrophoresis as a method of measurement of the ζ potential is suitable for particles of diameters below 1 μm. In particles of specific density close to the specific density of the solution, this range can be extended to a few micrometers. In particles of very high specific density (>5 g cm^{-3}), even dispersions of particles of diameters of about 0.1 μm are unstable. Particles, which are too large for electrophoretic measurements, may be suitable for the determination of ζ potential by electroosmosis or streaming potential (current).

Primary particles undergo coagulation (aggregation) and the aggregates are less stable against sedimentation than primary particles due to their size. In hydrophobic colloids, the rate of aggregation depends on the ζ potential of particles. Particles of ζ potentials below 10 mV in absolute value are very unstable against aggregation. In dispersions of such particles, aggregates are readily formed and they settle down before the sample is loaded into a chamber, and the measurement can be carried out. Thus, electrophoresis is suitable for particles having ζ potentials above 10 mV in absolute value. The limiting absolute value of 10 mV is a typical value. Individual dispersions may differ in their stabilities against aggregation at the same ζ potential, and the aggregation rate depends on the nature of solid particles (their Hamaker constant), particle size, the solid-to-liquid ratio, and the ionic strength of the solution. Certain dispersions are very unstable against aggregation in spite of a ζ potential as high as 20 mV in absolute value, while certain other dispersions are relatively stable in spite of a ζ potential as low as 5 mV in absolute value. Figure 2.20 shows that in principle ζ potentials high in absolute value are expected at low ionic strengths. Thus, electrophoresis is suitable for systems, in which the electrolyte concentration does not exceed 0.1 M.

The solid particles must not be too large to avoid sedimentation, but they must not be too small either. Very small particles (below 10 nm in diameter) may not scatter enough light to be visible. The lower limit of the size of the particles depends on their refractive index. Fine particles of high refractive index scatter more light and they may be better visible than larger particles, which have a refractive index matching the refractive index of solution.

The dispersion used in electrophoretic measurements has to be transparent; otherwise, the velocity of particles cannot be measured. The optical transparency of the dispersion depends on the concentration of particles, their size, and their refractive index. The electrophoretic measurements are carried out in dilute dispersions typically containing 1–1000 ppm of solid particles. A small amount of solid (<1 mg), which is sufficient to perform a measurement, is a substantial advantage of electrophoresis over the other electrokinetic techniques. Modern zetameters produce special messages when the concentration of particles is either too high or too low. Certain concentration of particles is necessary to record sufficient number of velocities of individual particles within a reasonable time. Moreover, in very dilute dispersions, particles, which are present in the dispersion as impurities (e.g., dust), may be more than the particles of interest, whose ζ potential has to be determined.

Certain commercially important dispersions may require adjustment of the solid-to-liquid ratio for electrophoretic measurements. It is not advisable for diluting such dispersions with water. When dispersion is too concentrated, a portion of the supernatant (obtained by filtration or centrifugation) can be used to dilute it without changing the composition of the solution (cf. the "Description of the system" section).

Other limitations and conditions
To calculate μ from Equation 2.26, the field strength E has to be known. When the solution is sufficiently conductive, a uniform electric field is

produced in the volume between the electrodes (Figure 2.21), that is, the lines of electric field are parallel. In such a system, E can be calculated as the quotient of the applied voltage and the distance between the electrodes. When the conductivity of the solution is too low, the lines of the electric field bend and the field is not uniform. Under such conditions, the field strength used in calculations by the instrument software may be different from the actual field strength, and the calculated μ is likely to be erroneous. This problem is especially significant when measurements in nonaqueous solvents are carried out in an experimental setup designed for aqueous systems. A special geometry of the measurement cell (designed to produce uniform electric field) allows measurements in low-conducting solutions.

The electric field in a measurement cell shown in Figure 2.21 induces an electroosmotic flow of solution near the walls of the container, which are electrically charged (Table 2.1). The velocity and direction of the electroosmotic flow depends on the sign and magnitude of the surface charge carried by the walls of the container. The flow of the solution near the walls of the container induces circulation of the solution in the entire container, which indirectly affects the observed velocity of the colloidal particles. The solution has to be kept in a container, hence electroosmotic flow is practically unavoidable in electrophoretic measurements. For common geometries of electrophoretic cells (cylinder, a slit between two parallel plates), analytical solutions for the velocity of the electroosmotic flow (as a function of the position in the cell) are readily available. In spite of the electroosmotic flow, which affects the observed velocity of particles, the actual electrophoretic mobility can be measured. This can be achieved either by measuring the electrophoretic mobility in certain regions of the cell where the solution is stagnant (no electroosmotic flow) or by correction of the raw (apparent) velocity of particles for the velocity of solution induced by electroosmosis. In certain types of zetameters, a special calibration of the optical system before the electrophoretic measurements is required to avoid errors due to electroosmotic flow.

Calculation of the ζ potential

Commercial zetameters use the following Smoluchowski equation:

$$\zeta = \frac{\mu\eta}{\varepsilon} \tag{2.27}$$

to calculate the ζ potential from the measured electrophoretic mobility (μ), where η and ε are viscosity and dielectric constant of the solution, respectively. In water at 25°C, Equation 2.27 produces a ζ potential of 12.8 mV for a typical μ of 10^{-8} m^2 V^{-1} s^{-1}. The viscosity of the solution is a function of temperature. Many modern zetameters have built-in thermostats, which allow for adjustment of the temperature of the sample. In case the temperature of the sample cannot be adjusted, it should be at least be controlled (also during the measurement) to assure that the proper value of viscosity is used in the calculations.

The Smoluchowski equation is valid for large particles and/or high ionic strengths. More precisely, it is a good approximation for $\kappa a > 100$, especially when the ζ potential is low in absolute value (30 mV or less), where κ and a are the reciprocal Debye length and the radius of a sphere (equivalent to the particle), respectively. Definition of an equivalent sphere in a dispersion of particles of different shapes and sizes is a problem by itself, and the discussion of this problem is beyond the scope of this section. For particles of 10 nm in radius, the Smoluchowski equation is valid for at least 10 M 1–1 electrolyte. For particles of 100 nm in radius, the Smoluchowski equation is valid for at least 0.1 M 1–1 electrolyte. For particles of 1 μm in radius, the Smoluchowski equation is valid for at least 0.001 M 1–1 electrolyte. There is an obvious contradiction between the above applicability ranges of the Smoluchowski equation, and the optimum conditions for electrophoretic measurements outlined in the "Systems Suitable for Electrophoretic Measurements" section. Namely, dispersions of fine particles (100 nm or less) in the range of applicability of Equation 2.27 are unstable because of fast coagulation (high ionic strength) and dispersions of larger particles (~1 μm) are unstable because of fast sedimentation of primary particles. Most dispersions for which actual electrophoretic measurements are carried out are beyond the range of applicability of Equation 2.27, and the ζ potential displayed by the instrument software is only an estimation. For $\kappa a < 100$, μ is a complicated function of ζ, κa, and of parameters characterizing the ionic composition of solution, and exact solutions are only available for highly symmetrical (e.g., spherical) colloidal particles. For irregularly shaped particles, such exact solutions do not exist. This is especially important to note that particles of different shapes and sizes have various μ at the same ζ. In general, Equation 2.27 underestimates the ζ potential of fine particles at low ionic strengths.

The exact relationship between μ and ζ for spherical particles (valid for any κa) was provided by O'Brien and White (1978). Calculations based on their model are tedious, and several simplified equations (valid for $\kappa a > 6$) have been forwarded. These equations may produce good approximations in dispersions of spherical "monodispersed" particles. In dispersions containing particles of different sizes, there is a problem, how to determine the representative κa to be used in the calculations.

In view of the difficulties discussed above in the calculation of the exact value of ζ potential from μ, application of Equation 2.27, even at $\kappa a < 100$ is understandable. Alternatively, many investigators report μ rather than the ζ potential to emphasize that in fact the ζ potential cannot be exactly calculated in dispersions containing irregularly shaped particles of various sizes. In dispersions of spherical "monodispersed" particles (or nearly spherical particles with a narrow distribution of particle size), application of the exact O'Brien and White theory is preferred rather than Equation 2.27.

Experimental difficulties

The electrophoretic measurements are carried out at a low solid load (often in a range of a few ppm). The surface area of solid particles exposed to the solution is low, and minute amounts of solutes may substantially change the surface. This is especially important in instruments, in which the same cell (Figure 2.21) is used for a series of various samples. Leftovers of surface-active substances from previous samples may affect the values of ζ potential observed in the current sample. Disposable cells solve this problem. In instruments, for which such disposable cells are not available, careful cleaning of the cell between the measurements is crucial. The choice of the cleaning agents depends on the character of the impurities to be removed and on the resistance of the cell to corrosive chemicals.

Sample preparation may also result in surface contamination at a low solid load. For example, silica (a component of glass) shows substantial solubility in water (~0.01 M). Storage of dispersions of certain metal oxides in glass containers, especially at neutral and basic pH, results in the adsorption of silicate anions (leached out from glass) on their surfaces. Thus, the surface acquires a negative charge and the ζ potential measured in a dispersion, which has been stored in glass, may be more negative than the ζ potential measured in the original dispersion. This problem is not unique for glass. Certain plastic containers contain water-soluble and surface-active additives, which affect the ζ potential of specimens stored in such containers.

Certain dispersions undergo coagulation and sedimentation on storage. Simple shaking is often not sufficient to destroy the aggregates. Ultrasonic treatment may help in re-dispersing the primary particles.

The electric current produces heat in electrolyte solutions. This effect is negligible at low ionic strengths and low voltages, but it becomes significant at higher ionic strengths. The local production of heat induces convective currents in solution, which affect the measured velocity of particles. This is an additional reason why electrophoretic measurements at high ionic strengths are not advised. When a high ionic strength is unavoidable, the time with electric field on should be minimized, and the voltage should be minimized, too, to avoid excessive production of heat. In certain types of zetameters, the instrument software will automatically prevent too high voltage and too high measurement time in the presence of highly conductive electrolytes. Proper choice of experimental conditions is a difficult compromise. Namely high voltage and long measurement time are desired in measurements of ζ potentials low in absolute value, which typically occur at high ionic strengths.

The electric current results in electrolysis of the electrolyte solution. The electrolysis products may be surface-active. They adsorb on solid particles and affect their ζ potential. This effect is negligible at low ionic strengths, but it becomes significant at higher ionic strengths. The adverse effects of electrolysis can be avoided by similar means as the heat production discussed in the previous paragraph. In addition, the electrophoretic cell can be designed in such a way that the portion of dispersion, in which the velocity of particles is measured, is separated from the near-electrode space (where the electrolysis occurs) by a thick layer of dispersion, thus minimizing the effect of adsorption of electrolysis products on the measured velocity.

The electric current results in polarization of the electrodes. Chemical changes in the electrodes themselves and in solution near the electrodes affect the field strength, which in turn affects the measured velocity of the particles. The electrode materials and the geometry of the cell are especially designed in commercial zetameters to minimize the electrode polarization. Polarization is negligible at low voltages and short measurement times, and it becomes significant at higher voltages and long measurement times. Proper choice of experimental conditions is a difficult compromise. Namely high voltage is desired to produce high (measurable) velocity of particles and long measurement times are desired to collect a substantial number of velocities of individual particles. In modern zetameters, the polarity of electrodes is alternated every few seconds (the particles and ions travel "back and forth") to minimize the electrode polarization. Special handling of the surface of the electrodes between series of measurements is advised in certain types of zetameters to minimize the effect of electrode polarization.

Electroosmosis and streaming potential (current)
Electroosmosis and streaming potential (current) measurements can be carried out for the same types of specimens and in a similar experimental setup. Certain instruments are so designed that they allow measurements by means of both techniques.

The experimental setup
The experimental setup used in electroosmosis is illustrated in Figure 2.22. The cell is filled with solution.

It is not much different from the experimental setup for electrophoresis presented in Figure 2.21 except that the cell contains solution rather than dispersion, and allows for a net flow of the solution while in electrophoretic measurements there are local flows of the solution, but they result in circulation rather than in net flow. In Figure 2.22, the specimen is also

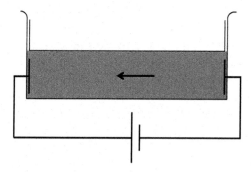

Figure 2.22 Electroosmosis. Specimen in the form of a capillary or a pair of parallel plates.

a cell (sample holder). This design has an advantage that the electric field does not interact with any electrically charged surfaces other than that of the specimen. Unfortunately, the properties and the amount of available materials seldom allow of making a capillary or a pair of parallel plates from a material of interest. Specimens in the form of porous monoliths or packed plugs (composed of fibers or particles of various shapes and sizes) can be studied by electroosmosis in an experimental setup shown in Figure 2.23.

The electroosmotic flow shown in Figure 2.23 results from the electric charge carried by the material of interest (sample) and by the sample holder. The cells in commercial zetameters are designed to minimize the effect of the sample holder on the measured quantities. In the streaming potential (current) technique, the experimental setup is similar to those shown in Figures 2.22 and 2.23 except that the potentiostat is replaced by a voltmeter, and the pressure gauge is complemented by a pump.

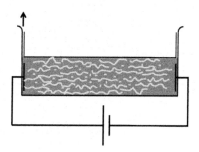

Figure 2.23 Electroosmosis. Specimen in the form of porous monolith or packed plug.

Systems suitable for electroosmotic and streaming potential (current) measurements

The porous plug has to be densely packed in order to avoid motion of particles in the course of the experiment. The surface area of the specimen exposed to solution should be substantially higher than the surface area of the sample holder. Fine (sub-micron) particles should be avoided, because the resistance of plugs made of fine particles to flow of the liquid is too high to allow the measurements. In this respect, electrophoresis and electroosmosis are complementary methods, that is, the former is suitable for fine, and the latter is suitable for larger particles. The resistance of a plug to flow of solution can be reduced by adjustment of its dimensions (thickness and cross-section), but such an adjustment is only possible to limited degree in commercial instruments.

Limitations, conditions, and experimental difficulties

Transparency to visible light and stability against sedimentation and coagulation, which define the applicability of electrophoresis, are not prerequisites in electroosmosis and streaming potential (current) measurements. Most other limitations, conditions, and experimental difficulties discussed in sections "Other Limitations and Conditions and Experimental Difficulties" for electrophoresis also apply to electroosmosis and streaming potential (current). The exposed surface area of the studied solid in electroosmosis and streaming potential (current) techniques can even be lower than in electrophoresis, especially when the specimen has a form of a capillary or a pair of parallel plates (Figure 2.22). Thus, the effects of minor amounts of impurities on the apparent ζ potential determined by electroosmosis and streaming potential (current) may even be more substantial than in electrophoresis. The problems with heat production and electrode polarization (at high ionic strengths) and with nonuniform electric field (at low ionic strengths) apply also to electroosmosis and streaming potential (current) techniques. The electroosmotic flow induced by a container (sample holder) may affect the apparent ζ potential in packed plugs (Figure 2.23), when the exposed surface area of the specimen is low, which may be the case in very coarse particles or macroscopic fibers.

Calculation of the ζ potential

The Smoluchowski equation for electroosmosis and streaming current assumes the following form:

$$\zeta = \frac{Il\eta}{\varepsilon S \Delta P} \tag{2.28}$$

where I is the electric (streaming) current in the cell, l is the length, S is the cross-section of the cell, and ΔP is the difference in pressure at the both ends of the cell. In electroosmosis, I is applied and ΔP is induced, while in streaming potential (current) ΔP is applied and I is induced. The ratio $I/\Delta P$ remains constant in a series of measurements at various applied I or ΔP.

Equation 2.28 can be re-written in various forms, for example, since:

$$I = \frac{U}{R} \tag{2.29}$$

where U is the electric (streaming) potential and R is the electric resistance of the solution, we have

$$\zeta = \frac{U I \eta}{\varepsilon S R \Delta P} \tag{2.30}$$

since the resistivity (of the solution):

$$\rho = \frac{RS}{l} \tag{2.31}$$

we have

$$\zeta = \frac{U \eta}{\varepsilon \rho \Delta P} \tag{2.32}$$

Electroacoustic techniques

The high-frequency (MHz range) analogs of classical electrokinetic phenomena are called electroacoustic phenomena, and commercial instruments based on such phenomena have been designed to determine the ζ potential in "fixed solution—moving particles" geometry (Table 2.1). Electroacoustic effect is observed when applied electric field induces a vibration of colloidal particles. Colloid vibration potential (current) is observed when applied acoustic wave induces an alternated electric field. Both techniques are applicable for dispersions of fine particles, and they are not applicable for macroscopic specimens, fibers, etc. Stability against sedimentation and coagulation, which define the applicability of electrophoresis are not prerequisites in electroacoustic techniques. Namely, dispersion is stirred and/or pumped during the experiment to prevent their sedimentation. The dispersion does not have to be transparent to visible light.

The signal is nearly proportional to the solid load, and a solid load of 0.5% to 20% is recommended in the ζ potential measurements. The signal may be too weak in more diluted dispersions to precisely determine the ζ potential. In this respect, electrophoresis and electroacoustics are complementary methods, that is, the former is suitable for diluted, and the latter is suitable for more concentrated dispersions. The above range of solid loads may be considered as an advantage, namely certain practically important dispersions have a solid load of 0.5–20%, and thus they may be studied by electroacoustic method directly, without additional sample preparation. In certain specimens, the above range of solid loads may be considered as a disadvantage, namely large amount of solid (typically >1 g) is required to perform a measurement, which may be a problem in certain materials. Ions present in solution produce a signal in electroacoustic techniques (even in the absence of any colloidal particles), which depends on the nature of the salt, and which is nearly proportional to the salt concentration. The raw signal produced by the instrument is the sum of the signal of ions and that of colloidal particles. In this respect, the electroacoustic techniques are completely different from classical electrokinetic techniques, in which the presence of charged surface is a prerequisite to obtain any signal. Several methods can be suggested to avoid errors in ζ potential measurements due to the signal produced by the ions. First, electrolytes, which produce strong signal (per unit of concentration), should be avoided. Certain electrolytes produce relatively weak signal even at high concentrations. The overall signal (ions + particles) is dominated by the signal produced by the particles when the solid load is high and the electrolyte concentration is low. Yet certain amount of electrolytes is necessary to keep the conductivity of dispersion at a desired level, to adjust the pH, etc. Commercial instruments are capable of collecting the signal of the electrolyte (in a separate measurement) and "subtracting" it from the signal of dispersion to produce "pure" signal of colloidal particles. This "electrolyte background subtraction" procedure offers a unique opportunity of ζ potential measurement in dispersions of particles in very concentrated electrolytes (up to 3 M), which are beyond the range of applicability of classical electrokinetic methods.

The signal is nearly proportional to the relative difference in specific density between the solid and the solution. In this respect, the electroacoustic techniques are completely different from classical electrokinetic techniques, in which the specific density of solid particles does not affect directly the measured signal. Relatively low ζ potentials can be measured by electroacoustic techniques in materials of very high specific density (e.g., zirconia) with sufficient precision, while the measurements in dispersion of solids of specific density nearly matching that of the solution (e.g., polystyrene latex) are not advised unless the particles are very highly charged. In this respect, electrophoresis and electroacoustics are complementary methods, that is, the former is suitable for low-specific-density solids (due to slow sedimentation of particles), whereas the latter is suitable for high-specific-density solids (due to strong signal).

The induced signal (electroacoustic wave or colloid vibration potential) shows certain inertia with respect to the applied field, and the phase shift can be used to determine the particle size in commercial instruments. Owing to high inertia, dispersions of very large particles (>10 μm) are not suitable for electroacoustic measurements, although the particles may still be kept in "dispersed" state due to a powerful stirrer.

References

Mishchuk, N.A. and P.V. Takhistov. Electroosmosis of the second kind. *Colloids and Surfaces A: Physicochemical and Engineering Aspects* 95(2–3), 1995: 119–31.
O'Brien, R.W. and L.R. White. Electrophoretic mobility of a spherical colloidal particle. *Journal of the Chemical Society, Faraday Transactions 2: Molecular and Chemical Physics* 74(0), 1978: 1607–26.

Further reading

Dukhin, A.S. and P.J. Goetz. *Ultrasound for Characterizing Colloids Particle Sizing, Zeta Potential Rheology.* Amsterdam, Netherlands: Elsevier Science, 2002.
Hunter, R.J. *Zeta Potential in Colloid Science: Principles and Applications.* London, UK: Academic Press, 1981.

Diffraction (XRD)

Deniz Rende

Introduction

This section is devoted to the understanding of structure of materials, especially, the arrangement of atoms in the solid state, using diffraction techniques. X-ray scattering and spectroscopy methods provide information on the physical and electronic structure of the crystalline and non-crystalline materials. X-ray diffraction is a powerful and rapid technique that requires comparatively easy sample preparation, fewer amounts of the sample ranging from a few milligrams to a few grams, and is non-destructive. However, its best practice is with homogeneous and single-crystal material and it generally requires an access to a standard reference file. X-ray diffraction has applications in:

- Characterization of crystalline materials
- Determination of unit cell dimension
- Measurement of sample purity

To use X-ray diffraction as a material's characterization tool, an understanding of diffraction theory and crystal structures is required.

Diffraction theory

Diffraction occurs when an electromagnetic wave comes across with regularly spaced obstacles that are able to scatter the wave and have spacings that are comparable in magnitude to the wavelength of the incident electromagnetic radiation (EMR). Waves are characterized by their wavelength (λ), which is defined as the distance between two consecutive peaks. Waves scatter from an object in all directions. When waves scatter from an object, they interfere with each other both constructively and destructively (Figure 2.24). When the two waves that have same wavelength, λ, and amplitude, A, are in phase after being scattered from an object, they interfere constructively resulting in a wave having the same wavelength but double the amplitude ($2A$); however, when the waves are out of phase after being scattered from an object, they interfere destructively, and cancel each other (Cullity and Graham, 2008; Callister and Rethwisch, 2012). If a second wave scatters from another object, displaced from the first by a distance in the order of the wavelength, then they can be viewed at some angle except at locations where they destructively interact.

X-ray diffraction is the elastic scattering of X-ray photons by atoms that are arranged in a periodic manner in space. X-rays have relatively short wavelengths with high energy (Lifshin, 2008). Owing to the wave nature of X-rays, the scattered X-rays from a sample can interfere with each other such that the intensity distribution is determined by the wavelength and the incident angle of the X-rays as well as by the atomic arrangement of the material, particularly the long-range order of crystalline structures. The expression of the space distribution of the scattered X-rays is referred to

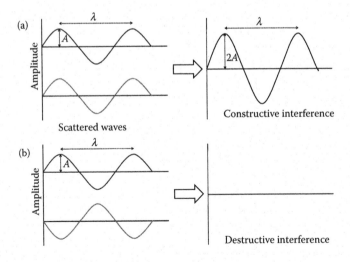

Figure 2.24 (a) Constructive and (b) destructive interferences of waves.

as the X-ray diffraction pattern. The atomic-level structure of the material can be determined by analyzing the X-ray diffraction pattern.

Crystal structures

Materials can be categorized in terms of the regularity with which their atoms are arranged with respect to each other. In a crystalline material, atoms are positioned in a repeating or periodic array over large atomic distances. The smallest possible repetitive structure that contains all the information to reproduce a crystal of any size is defined as the unit cell. Unit cells are parallelepipeds or prisms that are positioned around three interaxial angles. On a Cartesian coordinate system, the lengths of the sides (a, b, c) and the angles between the three sides (α, β, γ) are called the lattice parameters. For a cubic unit cell, $a = b = c$ and $\alpha = \beta = \gamma = 90°$.

To define crystal structures, the atoms are considered to be hard-spheres: the spheres touch each other and do not overlap. Four crystal structures are common for metals (Figure 2.25): simple cubic, face-centered cubic, body-centered cubic, and hexagonal close packed. Simple cubic (SC) model contains eight corner atoms. In this model, one unit cell includes one complete atom (one-eighth of each corner atom is inside the unit cell). In a face-centered cubic (FCC) unit cell, there are six half atoms located on the cube faces and eight corner atoms, therefore, an FCC unit cell includes four complete atoms. In a body-centered cubic (BCC) unit cell, a complete atom is located in the center of the unit cell in addition to the eight corner atoms, and therefore, two complete atoms fit into one BCC unit cell.

When dealing with crystalline materials, it is often necessary to describe certain features, such as a point, a direction, or a plane (Figure 2.26). Miller indices identify these features by using three integers: each corresponds to the axes on the Cartesian coordinate system. The points are defined by rational numbers, which are between 0 and 1, where brackets are used to show directions, and parentheses are used to indicate planes.

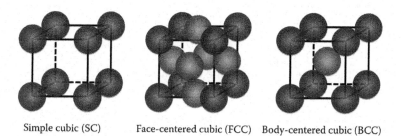

Simple cubic (SC)   Face-centered cubic (FCC)   Body-centered cubic (BCC)

Figure 2.25 Simple cubic, face-centered cubic, and body-centered cubic crystal models. (The atoms at the faces (FCC) and at the center (BCC) were colored for viewing purposes.)

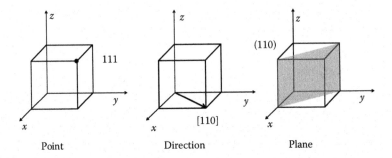

Figure 2.26 Illustration of a point, direction, and plane in a cubic unit cell.

In a cubic unit cell, a point is defined by the fractional multiples of the unit cell edges along the x, y, and z-axes. The position of the point is determined by the projection of this particular point to each axis. A crystallographic direction is the vector between two points in a crystal lattice. When defining, the vectors are translated to the origin (where the end of the vector is placed on 000), and the coordinates of the tip (arrow) indicate the point. The point coordinates should be integers, so any fractions should be cleared by multiplication with the lowest common factor. It is also necessary to define the planes in the crystal structure. The numbers used to label the planes are called the Miller indices, represented by (*hkl*). The parentheses in this notation are required and indicate that the *hkl* refer to a plane. A crystallographic plane either intersects or is parallel to at least one of the axes; the length of the intercept with its corresponding axis is used to determine the indices. The reciprocals of these values are multiplied with a common factor to obtain smallest possible integers (Messler, 2010).

From a crystallographic plane, it is possible to determine the spacing between similar planes or the lattice spacing. Figure 2.27 shows some plane examples. The distance between similar planes (d_{hkl}) are related to the Miller indices for a cubic lattice as follows:

$$d_{hkl} = \frac{a}{\sqrt{h^2 + k^2 + l^2}} \tag{2.33}$$

where a is the lattice constant of the cubic unit cell and h, k, and l are the Miller indices.

Bragg's law

The scattering process leading to diffraction can be equally visualized as if the X-rays are reflecting from imaginary planes defined by Miller indices. The X-rays hitting on a set of atomic planes, with indices *hkl*, make an

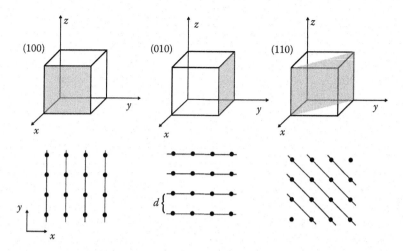

Figure 2.27 Some crystallographic planes through a cubic unit cell. The spacing between two planes is labeled as "*d*." The Miller indices for the shaded planes are shown in parentheses.

angle, θ, with them (Figure 2.28). The distance between the planes is d_{hkl}. Taking a close look at Figure 2.28, it can be seen that beam 2 travels more than beam 1 and the extra distance is given by AB + BC. If beams 1 and 2 simultaneously start in-phase, meaning that their waves are lined up peak to peak, then the extra distance (AB + BC) might cause beam 2 to be mis-aligned with beam 1 after reflection. The degree of misalignment is equal to the distance (AB + BC). Constructive interference occurs when AB + BC equals integer multiples of the wavelength of the beam:

$$n\lambda = AB + BC \qquad (2.34)$$

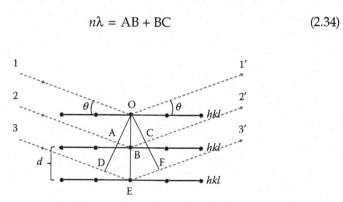

Figure 2.28 Diffraction of X-rays from consecutive planes in a crystal. Red dots represent the locations of the atoms in the crystal. The dark lines connecting the atoms are drawn to indicate the planes formed by the atoms.

Table 2.2 Selection Rules for Reflections in Cubic Crystals

	$h+k+l$	SC	BCC	FCC
(100)	1	✔	✘	✘
(110)	2	✔	✔	✘
(111)	3	✔	✘	✔
(200)	2	✔	✔	✔
(210)	3	✔	✘	✘
(211)	4	✔	✔	✘
(220)	4	✔	✔	✔

where n is a positive integer. The distance (AB + BC) can be measured in terms of the angle θ by considering the triangle $\triangle ABO$ and $d \sin \theta = AB$, or $2d \sin \theta = AB + BC$. Then the condition for diffraction to occur is

$$n\lambda = 2d_{hkl} \sin \theta \tag{2.35}$$

This equation is called the Bragg's law and allows one to relate the distance between the planes of a crystal to the diffraction angle and X-ray wavelength. When the wavelength, λ, is known, and the angle, θ, is measured, the interatomic distance, d_{hkl}, can be calculated for a specified unit cell.

Certain cubic crystal structures create constructive interference at certain planes. For instance, it is possible to obtain constructive interference for all the planes from a simple cubic cell. However, a body-centered cubic cell allows reflections only when the summation of the Miller indices $(h + k + l)$ for the particular plane is equal to an even integer. When all individual Miller indices are even or odd, then a face-centered cubic cell allows for the reflection on this plane. These rules are summarized in Table 2.2.

Sample preparation and instrumentation

The mechanical assembly of X-ray diffractometers consists of three basic elements: an X-ray tube (source), a sample holder, and an X-ray detector. When the sample is tilted with an angle θ, while a detector rotates around it on an arm at twice this angle, this system is known as the Bragg–Brentano system (Clearfield et al., 2008). The working principle of a Bragg–Brentano system is given in Figure 2.29. In this assembly, the distance from the X-ray focal spot to the sample is the same as that from the sample to the detector.

Sample holder and sample preparation

The most important part of the X-ray diffraction analysis is the sample preparation. While preparing for an X-ray analysis, it is important to have

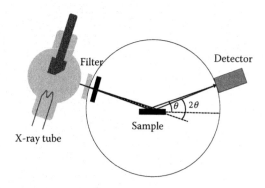

Figure 2.29 Illustration of Bragg–Brentano system.

a specimen that is representative of the sample, hence by definition, a sample is the material to be tested, whereas a specimen is relatively a small portion of the sample to be tested. Considering majority of the samples are in powder form, or at least retain their solid state, only preparation of powder samples are explained in detail.

Powder, by definition, is a solid state containing small particles. A particle in this case could be a single crystalline, polycrystalline, or an amorphous solid, where only crystalline powders exhibit diffraction. In powder or polycrystalline diffraction, the sample surface should be flat and homogeneous, as well as the crystals should be randomly distributed. Only crystallites having reflecting planes (h, k, l) parallel to the specimen surface will contribute to the reflected intensities. If we have a truly random sample, each possible reflection from a given set of h, k, l planes will have an equal number of crystallites contributing to it. Hence, it is desired to reduce the particle size to orient the crystals randomly, rather than in a preferred orientation to have a complete representation of the specimen and enhance the statistical significance of the analysis (related to the randomness of the specimen). A few grams of the sample is ground with a mortar and pestle to a fine powder, typically of size ~10 μm. The best practice is to grind the sample by making small circles with the pestle while pressing it against the mortar for about 15 min. Agate mortar and pestles are preferred to grind samples; ceramic mortar and pestles are prone to contamination. When the sample amount is too less and the material is soft, one other alternative to grind the powder is to place it between two microscope slides and to twist the slides slowly.

When a sample holder is used, the volume of the holder should be enough to hold the entire powder sample (Figure 2.30). It is essential to generate a flat surface; therefore, the sample is placed in the holder well until it spills over onto the top. Then the powder is pressed with a glass slide and excess material is removed by sliding the glass slide over the material.

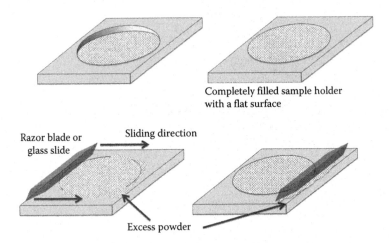

Completely filled sample holder with a flat surface

Razor blade or glass slide

Sliding direction

Excess powder

Figure 2.30 Preparation of a powder sample for X-ray diffraction.

Dusting in oil, grease, or silicon and mixing it with an inert powder (wheat) and volatile inert liquids (acetone, ethanol) are alternatives if the amount of the powder sample is not enough to fill the sample holder. Then the sample is placed in a sample holder or onto a holder surface either by smearing on a glass slide while making a flat surface or sprinkling on a double-sided tape.

Another option to prepare specimens is the spray-drying, which is basically suspending the powder in water and spraying the suspension in a heated chamber. The instrument involves a heating chamber located vertically, where an airbrush is located at the top. When the ground particles suspended in water is sprayed into the chamber, the particles dried in the chamber fall down with gravity to be collected on a tray or a sheet of paper (Figure 2.31) (Hillier, 1999). The resulting dry product consists of thousands of tiny spherical granules of the sample components. Typically, the average diameter of the granules is about 50 μm (Hillier, 2002). One of the drawbacks of this method is the amount of material that is lost during the spraying process. Also, the use of water as the suspension medium restricts the type of materials: only some are dissolved in water.

Source

X-rays are generated in a cathode tube by heating a filament to produce electrons, accelerating electrons toward a target by applying voltage, and bombarding the target material with electrons. When electrons have enough energy to remove inner shell electrons, the characteristic X-ray spectra are produced.

The production of X-rays generates large quantities of heat, which must be dissipated rapidly in order to prevent metallic targets from melting,

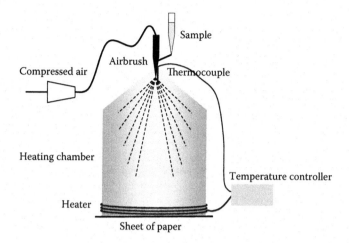

Figure 2.31 Illustration of a spray-dryer used to prepare X-ray diffraction powder samples.

therefore durable, heat- and electrical-conductive metals are used as generators. Most laboratories use a sealed X-ray tube with a target of copper, cobalt, chromium, and molybdenum. These metals produce X-rays in the 4–21 eV range and provide stable heat conduction and corrosion resistance. Upon the removal of electrons from K, L, and M orbitals, representing the lowest energy orbitals surrounding the nucleus, an electron from a higher energy orbital instantaneously falls to the lower energy orbital, while a photon of high energy is emitted. The energy of the photon is fixed and known as the characteristic wavelength. The majority of the X-rays are Kα line (λ = 1.54 Å for Cu, λ = 0.7 Å for Mo), which is the transition from the L shell to the K shell. Only two electronic transitions from the L shell to the K shell are allowed, these transitions split the Kα radiation, as Kα1 and Kα2, where Kα1 has a slightly shorter wavelength and twice the intensity as Kα2. For instance, Kα line is a doublet (1.54051 and 1.54433 Å for Cu) radiation, which requires a monochromator to pass only one component. To produce a monochromatic X-ray, the radiation is filtered through a single crystal placed either adjacent to the source, that is, primary monochromator or just before the detector, known as diffracted beam monochromator.

The incident X-ray causes fluorescent radiation of the sample excited. The high-energy X-rays knock out the electrons in inner shells, the electrons in the outer shells drop down to occupy empty levels while excess energy is emitted as fluorescence, which acts as a secondary X-ray source. Almost all elements emit fluorescent radiation but its effect is more apparent when the energy of the fluorescent radiation is close to that of the characteristic radiation used. This characteristic radiation from the sample is accepted by the X-ray detection system, which generates a background in

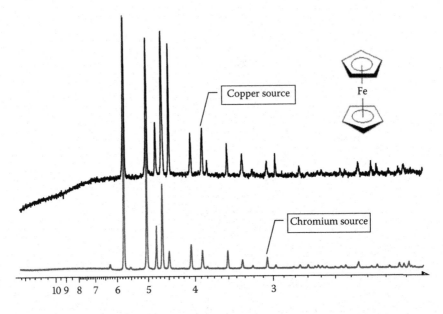

Figure 2.32 X-ray diffraction pattern for ferrocene (FeCp2) recorded by using two different sources. (From Clearfield, A., J. Reibenspies, and N. Bhuvanesh. *Principles and Applications of Powder Diffraction.* Wiley, 2008. With permission.)

the final X-ray diffraction pattern. For instance, when the X-rays generated by the copper source is targeted to iron or cobalt specimens, they fluoresce strongly. Hence, the selection of a proper X-ray source is important and should be chosen carefully based on the material to be tested. Depending on the source, spectra could be enhanced. Figure 2.32 shows the X-ray diffraction of ferrocene (FeCp2) when chromium source is used instead of copper. The intensity values are represented as arbitrary values, however the diffracted intensity for Cr radiation is ~15 times higher than that of Cu and the detectors that collect the diffractions are different. However, the background X-ray fluorescence is obvious when copper source is used (Clearfield et al., 2008).

Detector

The selection of the X-ray detector also determines the quality of the analysis. The working principle of detectors relies on the interaction of the X-ray photon with an atom in the detector, which is excited and counted upon returning to ground state. If the excited atom produces photon (light), then the irradiation is captured by a CCD camera (Clearfield et al., 2008). There are a large number of detectors suitable for powder X-ray diffraction. The simplest is the photographic film, which allows for the collection of entire diffractogram at one time. Careful handling of the photographic

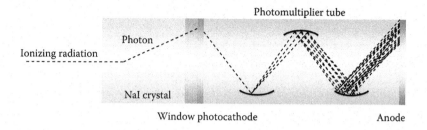

Figure 2.33 Illustration of a scintillation counter.

film may allow for the quantification of intensities. The well-known detectors are scintillation, proportional, and semiconductor.

The most widely used detector is the scintillation counter, in which the X-rays are converted into visible light, typically in a thallium-doped sodium iodine (NaI) crystal (Figure 2.33). The energy of the visible light produces electrons. The number of electrons is multiplied by a series of metal plates until 1–10 V of electric current is generated.

The proportional detector is a metal container with an X-ray transparent window, where the container is filled with an inert gas (Figure 2.34). The set-up is completed with a wire, which serves as an anode, whereas the container itself acts as a cathode. When an X-ray photon enters through the window, it interacts with an inert gas atom ejecting an electron from the gas atom. While the ejected electrons travel to the anode, the ionized atoms move to the cathode: this generates an electric current, where the electric potential is recorded. The signal produced in this assembly is

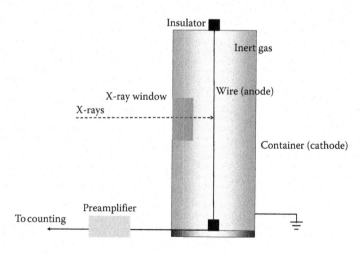

Figure 2.34 Illustration of a proportional detector.

proportional to the energy of the X-ray photon that hits the detector, hence the name is proportional.

Various semiconductor detectors (Si:Li, positive–intrinsic–negative (PIN)) offer energy resolutions up to 100–300 eV, sufficient to distinguish fluorescence from different elements and from diffracted X rays.

Analysis of X-ray diffraction

The useful powder diffraction data set includes an X-ray beam impinged on a sample and recordings of the diffracted intensity as a function of the angle. A simple representation of the set-up is presented in Figure 2.35. In a typical experiment, monochromatic (single wavelength) X-ray is directed on the sample, as the sample or the detector rotates, the diffraction measurement is recorded as 2θ with a detector. The measurements are generally optimized on intensity, resolution, and elimination of undesired effects (background from sample fluorescence).

To assign the observed peaks to the crystallographic planes, either the diffraction pattern can be compared with existing libraries or the intensity peaks are assigned by following simple rules. Today about 50,000 inorganic and 25,000 organic single component, crystalline diffraction patterns have been collected in databases. International Centre for Diffraction Data (www.icdd.com), Inorganic Crystal Structure Database (http://www.fiz-karlsruhe.de/icsd-content.html), and International Union of Crystallography (http://www.iucr.org) offer generous X-ray diffraction data repositories and software for analyzing crystallographic measurements.

Assignment of the intensity peaks to crystallographic planes can be accomplished by following simple rules. Consider that the X-ray measurement gives you the 2θ angles. To find the *hkl* indices for each peak, the

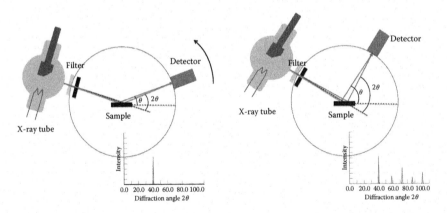

Figure 2.35 The collection of the X-ray diffraction data.

link between the Bragg's law to the *hkl* indices should be established. The interplanar spacing, d_{hkl}, serves as a connection as shown in the following equations (for cubic crystals only):

$$\lambda = 2d_{hkl} \sin \theta. \tag{2.36}$$

The relation of the interplanar spacing to the Miller indices is

$$d_{hkl} = \frac{a}{\sqrt{h^2 + k^2 + l^2}}. \tag{2.37}$$

Rearranging these two equations gives

$$h^2 + k^2 + l^2 = \left(\frac{4a^2}{\lambda^2}\right) \sin^2 \theta. \tag{2.38}$$

Recalling that *h*, *k*, and *l* values are all integers, therefore, $h^2 + k^2 + l^2$ must be a positive integer. For a monochromatic X-ray (hence λ is fixed) and a cubic crystal structure, the following rules could be applied to the experimentally determined X-ray peaks such as the one shown in Figure 2.36:

1. Calculate the $\sin^2 \theta$ values for each peak.
2. Normalize the $\sin^2 \theta$ values with the first peak's $\sin^2 \theta$ value.
3. Multiply the normalized values with a common multiplier to convert them into integers. These integer values are equal to $h^2 + k^2 + l^2$.

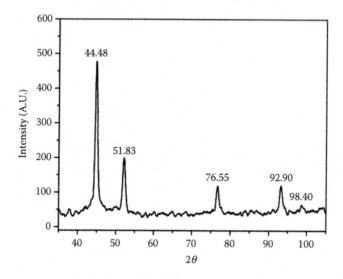

Figure 2.36 X-ray diffraction pattern for iron.

Table 2.3 Determination of the Cubic Unit Cell Structure

2θ	$\sin^2 \theta$ (step 1)	Normalized values (step 2)	Convert to integers (step 3)	Predicted h, k, l (step 4)	Verify $\lambda^2/4a^2$ step (5)
44.48	0.143	1	3	(111)	0.0477
51.83	0.191	1.34	4	(200)	0.0477
76.55	0.383	2.68	8	(220)	0.0479
92.90	0.525	3.67	11	(311)	0.0477
98.40	0.573	4	12	(222)	0.0478

4. One can now predict individual h, k, and l values such that the sum of the squares of the predicted values is equal to the integers calculated in step 3.
5. Verify that $4a^2/\lambda^2$ (or its reciprocal) is the same for each predicted h, k, and l value.

This procedure can be understood well with an example. Figure 2.36 shows X-ray diffraction data for iron generated by a copper source with a wavelength of 0.154 nm (Cu Kα). The crystal structure (face-centered cubic or body-centered cubic) and lattice constant, a, can be calculated. When the lattice constant and the crystal structure of a material is known, then the radius of an atom can be calculated, which helps one to identify the material. The crystal structure (face centered cubic or body centered cubic) and lattice constant, a can be calculated. The steps are summarized in Table 2.3.

Notice that hkl are all odd or all even, therefore the crystal structure of this metal is face-centered cubic according to the selection rules given in Table 2.2. The last column also verifies that the $\sin^2 \theta/(h^2 + k^2 + l^2)$ values are constant. The lattice constant, a, can then be calculated (for a given wavelength, 0.154 nm) as

$$0.0477 = \frac{(0.154\,\text{nm})^2}{4a^2},$$

where $a = 0.352$ nm.

When the lattice constant value is combined with the crystal structure information, the distance between the adjacent atoms can be determined.

References

Callister, W.D. and D.G. Rethwisch. *Fundamentals of Materials Science and Engineering: An Integrated Approach*. New York: Wiley, 2012.

Clearfield, A., J. Reibenspies, and N. Bhuvanesh. *Principles and Applications of Powder Diffraction*. Chichester, UK: Wiley, 2008.

Cullity, B.D. and C.D. Graham. *Introduction to Magnetic Materials*. Hoboken, NJ: John Wiley & Sons, 2008.

Hillier, S. Spray drying for X-ray powder diffraction specimen preparation. *Commission on Powder Diffraction - International Union of Crystallography 27*, 2002: 7–9.

Hillier, S. Use of an air brush to spray dry samples for X-ray powder diffraction. *Clay Minerals 34*, 1999: 127–35.

Lifshin, E. *X-Ray Characterization of Materials*. Weinheim, Germany: Wiley, 2008.

Messler, R.W. *The Essence of Materials for Engineers*. Sudbury, MA: Jones & Bartlett Publ., 2010.

Further reading

Guinier, A. *X-Ray Diffraction: In Crystals, Imperfect Crystals, and Amorphous Bodies*. New York, NY: Dover, 1994.

Jenkins, R. and R.L. Snyder. *Introduction to X-Ray Powder Diffractometry*. New York, NY: Wiley, 1996.

Sibilia, J.P. *A Guide to Materials Characterization and Chemical Analysis*. VCH, 1996.

Stout, G.H. and L.H. Jensen. *X-Ray Structure Determination: A Practical Guide*. New York, NY: Wiley, 1989.

Scattering

Ulf Olsson

Introduction

Light that interacts with matter can be either absorbed and/or scattered. The latter is a prerequisite for our ability to see, and combined with absorption phenomena, a color vision. In a more general sense, there are essentially two ways by which we can "see," if by seeing we mean determining the size and the shape of the objects. The first, and the most commonly known way of seeing, may be referred to as imaging. Here, by the use of a lens, the light that is reflected or scattered from an object is focused to create an image of the object on a screen. This is, in principle, how our eyes work too. The light is focused by our eye lenses, sometimes in combination with glasses when the eye lenses are not operating properly, to create an image on the retina in the back of the eye, which is then analyzed by the brain. The second way of "seeing," may be the scattering or diffraction. Here, we work without focusing the lens and instead measure and analyze the scattered intensity as a function of the scattering angle, θ, as illustrated in Figure 2.37.

The scattered intensity, $I(q)$, is generally presented, not as a function of the scattering angle, but as a function of the so-called scattering vector (magnitude) $q = |\vec{q}|$, where $\vec{q} = \vec{k} - \vec{k}_0$ (see Figure 2.37) is the difference between the wave vectors of the scattered ($\vec{k}$) and incident ($\vec{k}_0$) beams,

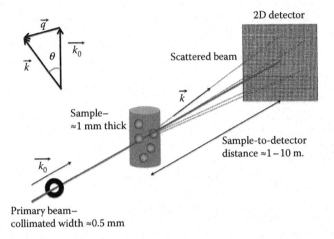

Figure 2.37 Schematic illustration of a (small-angle) scattering experiment. An incoming primary beam, with wave vector $\vec{k}_0$, impact on a sample and the scattered radiation (wave vector $\vec{k}$) is here recorded by a two-dimensional detector. The scattering vector, $\vec{q} = \vec{k} - \vec{k}_0$, is also illustrated.

respectively. If the scattering event is purely elastic, $|\vec{k}| = |\vec{k}_0| = 2\pi/\lambda$, where λ is the wavelength of the radiation, the scattering vector magnitude is related to the scattering angle through:

$$q = \frac{4\pi}{\lambda}\sin\frac{\theta}{2} \tag{2.39}$$

In a scattering or diffraction experiment,[*] a scattering or diffraction pattern is obtained that carries the information about the size and shape of the object, the number or concentration of objects, and how the objects are distributed in space. A common and simple demonstration of the phenomenon, which can be found in general physics textbooks, is the diffraction from a thin slit, illustrated in Figure 2.38. A laser beam passing through a thin slit gets diffracted, and a characteristic diffraction pattern can be obtained on a screen behind the slit. This pattern consists of intensity maxima and minima, where the exact pattern (angles or q-values of maxima and minima) depends on the slit width.

The radiation is scattered by atoms and/or molecules, where different atoms/molecules have different scattering power. This results in a scattering contrast between different molecules, as the one between a polymer

---

[*] The word diffraction is mainly used when the experiment is carried out on crystalline solids to obtain a crystalline structure, while the word scattering is mainly used for the analysis of the more diffuse scattering from liquids or amorphous objects.

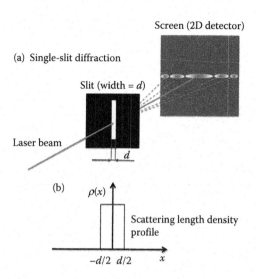

Figure 2.38 (a) Illustration of the single-slit diffraction experiment. Note that diffraction is mainly observed perpendicular to the thin slit. (b) The slit example corresponds to a rectangular scattering length density profile.

and a solvent, or between the domains of different composition as in an emulsion, or in an aerosol. The scattering pattern, $I_{sc}(q)$, is related to the real space structure through a Fourier transform. We may illustrate this by the single-slit diffraction in Figure 2.38, as the calculation here is relatively simple. Consider a long vertical, but narrow, slit. Because it is long in the vertical direction, it scatters mainly in the horizontal direction (as will be explained below) and we need to consider only one dimension, the x-direction. There is a finite scattering power only inside the slit. Outside the slit, the beam is blocked and scattering power is zero.

Denoting the scattering power by ρ, its variation in the x-direction, the scattering power profile $\rho(x)$, is illustrated in Figure 2.38b, where δ is the slit width. In this example, the profile is rectangular with $\rho(x) = \rho_0$ inside the slit, and $\rho(x) = 0$ elsewhere. The scattered intensity is now given by

$$I_{sc}(q) = \left| \int_{-\infty}^{\infty} dx \rho(x) e^{iqx} \right|^2 = \rho_0^2 \left| \int_{-d/2}^{d/2} dx\, e^{iqx} \right|^2 \tag{2.40}$$

This Fourier integral has a relatively simple solution. We recall that $e^{i\theta} = \cos\theta + i\sin\theta$. Since the integration interval is even, the integral over the imaginary part vanishes, as the sine function is an odd function. Equation 2.40 can thus be written as

$$I_{sc}(q) = \varrho_0^2 \left| \int\limits_{-d/2}^{d/2} dx \cos qx \right|^2 = 4\varrho_0^2 \left(\frac{\sin\{qd/2\}}{q} \right)^2. \qquad (2.41)$$

This particular scattering pattern has zero intensity at $q = 2\pi/d$, $4\pi/d$, $6\pi/d$, ... and intensity maxima at $q = 0$ and at $3\pi/d$, $5\pi/d$, $7\pi/d$, ..., and where the intensity of the maxima decays as q^{-2}. For example, by determining the q-value(s) where the intensity is zero or has a maximum, we can obtain the slit width. In practice, this is done by measuring the corresponding scattering angles and knowing the wavelength of the light source. We see from Equation 2.41, as d increases, the positions of the zeros move to lower q-values. In the limit of $d \to \infty$, all the zeros move into $q = 0$. This means that there is no scattering. Mathematically, this situation is described by $\int_{-\infty}^{\infty} dx\, e^{iqx} = \delta(0)$, where $\delta(x)$ is the Dirac delta function. Scattering only in the forward direction, $q = 0$, is equivalent to propagating a beam (no scattering), and that is why we essentially do not see any vertical scattering from the slit when the slit length is large.

There is a fundamental condition that needs to be fulfilled in order to be able to "see" the shape and size of the objects. This condition, which holds both for "seeing" by imaging or by scattering, is that the wavelength, λ, of the radiation used to "see" with, has to be smaller than the size of the object. This means that, no matter how much we magnify, in a light microscope we cannot see objects that are smaller than a micrometer, which is approximately the wavelength of visible light. In order to "see" smaller things, the wavelength has to be made shorter. This is essentially not a problem. Different sources with shorter wavelengths are available and there are also suitable detectors of such radiation, as we cannot use our eyes directly when we depart from the visible part of the spectrum. UV light, however, is not suitable for seeing. The reason is that essentially all matter absorbs rather strongly in UV, making all objects and matter dark. By decreasing the wavelength further into the so-called X-ray regime, absorption gets decreased again and matter becomes increasingly transparent. While there still remains significant absorption, X-rays with $\lambda \approx 0.1$ nm (1 Å) is commonly used in scattering experiments.

This fundamental condition is also illustrated in Figure 2.39. In order to determine the width of the slit, we need to measure the q-value of at least the first intensity minimum occurring at $q = 2\pi/d$. Since the maximum scattering angle is $180°$, observing the first minimum requires $\lambda < 2d$.

Besides X-rays, one can also produce electron and neutron beams with short wavelengths, suitable to see small things. For such particle beams, the wavelength is related to the momentum, $p = mv$, through $\lambda = h/p$, where m is the mass, v the velocity, and h is Planck's constant. A beam of charged electrons can be focused using electromagnetic lenses and are mainly used

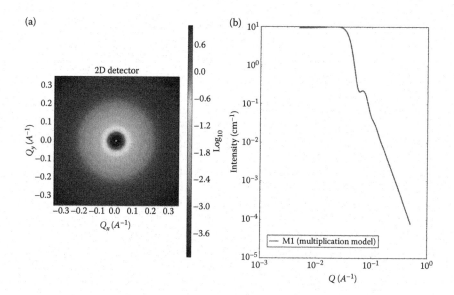

Figure 2.39 (a) Typical two-dimensional small-angle X-ray scattering pattern from a dispersion of spherical colloidal particles. (b) Radially averaged scattered intensity, $I(q)$, plotted as a function of q.

in electron microscopes. Neutrons can be produced with a suitable wavelength in reactors, or by the so-called spallation, where protons are accelerated to high velocities and then impact on heavy atoms. In this spallation event, many neutrons are released. Neutrons are mainly used for scattering experiments and have some particular advantages compared to X-rays. For most materials, absorption is negligible and neutrons can therefore penetrate deep into the materials. The scattering mechanism involves interactions with the atomic nuclei and the scattering power, and thus the contrast can be varied by isotopic substitution. In particular, substituting normal hydrogen with deuterium is often used to highlight certain parts of the molecule. On the other hand, X-rays are scattered by electrons and the atomic scattering power thus increases with the atomic number. Hydrogen, which has only one electron, is almost invisible to X-rays while it strongly scatters neutrons. Neutron scattering is therefore important for studies of hydrogen-storing materials and to refine protein structures when hydrogen positions are important. Neutrons, being spin = 1/2 particles, carry a magnetic moment and are also scattered by local magnetic fields inside materials which is used in the studies of high-temperature superconductors.

Above we have discussed scattering experiments in general terms. For what remains, we will focus on some particular scattering experiments that are commonly used to study colloids, gels, polymers, lipids proteins, or other soft matter systems, with characteristic (colloidal) length scales in the 1 nm to 1 μm range. These are small-angle scattering of X-rays or neutrons, and static and dynamic light scattering.

Small-angle scattering

Scattering experiments probe the structure on the length scale $2\pi/q$. Hence, in order to investigate structures on the colloidal length scale, we typically need access to the q-range 0.01–10 nm. With X-rays of wavelength 0.1 nm, this corresponds to small angles, 0.01–10°, as obtained using Equation 2.39. When we are interested only in small angles, in which a typical instrument design involves a narrowly collimated X-ray beam, produced by an X-ray source and giving a spot size of a fraction of a mm (≤0.5 mm) on a sample typically contained in a glass capillary of 1 mm in diameter. The scattered intensity is recorded by a stationary, typically a two-dimensional area detector (e.g., a CCD camera), placed behind the sample, as illustrated in Figure 2.37. On the detector, there is also a beamstop that protects the detector from the intense primary beam, and the whole or at least most of the system is kept under vacuum to avoid scattering from air that would result in a background noise. The available q-range depends on the diameter of the detector, the diameter of the beamstop, and the sample-to-detector distance. Reaching very low q-values requires a very long sample-to-detector distance (several meters) and a narrow beam and beamstop, while larger q-values can be obtained by shortening the sample-to-detector distance. To cover a large q-range, it is clearly advantageous to be able to vary the sample-to-detector distance. Alternatively, one sometimes uses two detectors simultaneously, one low-angle and one wide-angle detector. For a more detailed description, see, for example, the review of Narayanan (2009).

From isotropic solutions or dispersions, the scattering pattern is circularly symmetric (Figure 2.39a). This pattern is then usually radially averaged to produce the one-dimensional scattering pattern $I(q)$ versus q (Figure 2.39b). The intensity is a measure of the energy flux, that is, the number of photons/particles, passing a unit area perpendicular to the propagation per unit time. The scattered intensity, I_{sc}, recorded by the detector, depends on the intensity of the primary beam, I_0, on the irradiated sample volume, V_s, and on the sample-to-detector distance, l_{s-d}, parameters that vary from experiment to experiment. It is therefore useful to define an absolute intensity scale, where the experimentally recorded intensity is normalized with respect to these variables. I_{sc} is proportional to I_0 and V_s and because radiation intensities decay as the inverse square of the distance, the absolute scaled intensity is defined as

$$I = \frac{l_{s-d}^2}{I_0 V_s} I_{sc} \qquad (2.42)$$

In terms of the wave description of the radiation, $A(x) = A_0 \cos(x)$, the intensity is given by the square of the amplitude, $I = |A_0|^2$, which is the reason for the square in Equations 2.40 and 2.41. To obtain the experimental scattering data on absolute scale, one typically do a calibration with a known scatterer. In the case of X-rays, one often chooses pure water.

Turning now to the scattering from colloidal particles, we consider first a single spherical colloidal particle in vacuum. This can serve as an example of aerosols or the ice particles of the clouds in the sky, if we neglect the scattering from the air molecules. The single-particle scattering function is given by the square of a Fourier integral, as in the example of the slit in Equations 2.40 and 2.41. Now, however, we have to integrate over a sphere in three dimensions, with the position/coordinate vector $\vec{r} = (x, y, z)$. If $\rho(x) = \rho_p$ inside the sphere while being zero elsewhere $I(q)$ for the sphere is given by

$$I_{sc}(q) \sim \varrho^2 V_s^2 \left(\frac{3(\sin(qR)) - qR\cos(qR))}{(qR)^3} \right)^2 \tag{2.43}$$

where R is the sphere radius and $V_s = 4\pi R^3/3$ the sphere volume. If the particle surrounding is not vacuum but a solvent, then ρ in Equation 2.43 should be replaced by $\Delta\rho$, the difference in ρ between the particle and the solvent. Strictly, ρ has the dimension of length^{-2}, for example, cm^{-2}. It is often referred to as the scattering length density, and $\Delta\rho$ is the contrast between the particle and the surrounding solvent. If $\Delta\rho = 0$, the particle is "invisible." X-rays are scattered by individual electrons and for this type of radiation, ρ is proportional to the electron density of the material and X-ray contrast is obtained by differences in electron density.

The q dependence of Equation 2.43 is given by the function

$$P(q) = \left(\frac{3(\sin(qR) - qR\cos(qR))}{(qR)^3} \right)^2 \tag{2.44}$$

where $P(q)$ is referred to as the (normalized) particle formfactor, and it is this part that carries the information of the particle shape, here a homogeneous sphere. $P(q)$ is normalized, so that in the limit of $q = 0$, $P(0) = 1$. For cylindrical or disc-shaped objects, this function is different. Also, if the particle is not homogeneous, but, for example, is hollow, consisting of a spherical shell, as for a lipid vesicle, $P(q)$ is significantly different. A library of formfactors can be found in the review by Pedersen (1997).

$P(q)$ for homogeneous spheres of radii $R = 10$, 50, and 100 nm are presented in Figure 2.40, in a double-logarithmic plot. The normalized $P(q)$

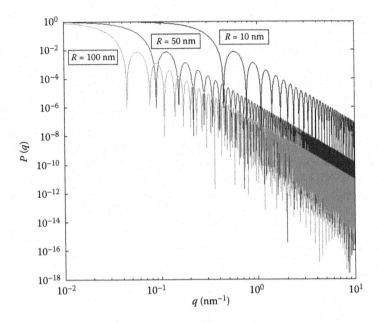

Figure 2.40 P(q) for homogeneous spheres of radii R = 10, 50, and 100 nm, respectively.

begins at 1 at $q = 0$ and then decays with increasing q, showing oscillations also due to the sine and cosine functions (Equation 2.44). The particular pattern of maxima and minima reports on the particle's shape. The alternating maxima and minima (zeros) shift to lower q with increasing R.

Typical scattering experiments are performed on solutions or dispersions with a large number of particles in the scattering volume. When they are dilute, with the average nearest-neighbor distance being much longer than R, the total scattering will just be the sum of the scattering from the different particles. At higher concentrations, however, the scattered intensity will also depend on the structure formed by the particles in the solution, a consequence of inter particle interactions. In contrast to crystals, however, where the structure is of long range, the structure in solutions is only of short range. The effect is accounted for by multiplying with the so-called structure factor, $S(q)$, so that our final expression for the absolute scaled scattered intensity is

$$I(q) = \frac{N_p}{V} V_p^2 \Delta \varrho^2 P(q) S(q) \qquad (2.45)$$

where N_p/V is the number density of particles, V_p is the particle volume, $P(q)$ is the normalized formfactor, and $S(q)$ is the structure factor. $P(q)$ and

$S(q)$ are dimensionless and we then see in Equation 2.45 that $I(q)$ has the dimension of inverse length.

In statistical mechanics, the structure of liquids is generally described in terms of the radial distribution function, $g(r)$, that tells us the probability of finding another particle (or the average particle concentration) at a distance r from a given test particle. $S(q)$ is the Fourier transform of $g(r)$ and thus carry the same information about the structure, but in the reciprocal space. From the value of structure factor in the limit of $q = 0$, we can obtain the osmotic compressibility of the sample:

$$S(0) = k_B T \left(V_p \frac{\partial \pi}{\partial \phi} \right)^{-1} \tag{2.46}$$

as a measure of the inter particle interactions. Here, π is the osmotic pressure and ϕ is the volume fraction of particles.

In Figure 2.41, we have plotted $S(q)$ as a function of qR for some different concentrations of spherical particles interacting as hard spheres. For this case, there exists an analytical expression that can be found in Kinning and Thomas (1984). At low concentrations, the particles are essentially non-interacting and the solutions behave as ideal with $\pi = \phi k_B T / V_p$, and $S(q) \approx 1$ for all q-values. With increasing concentration, the excluded

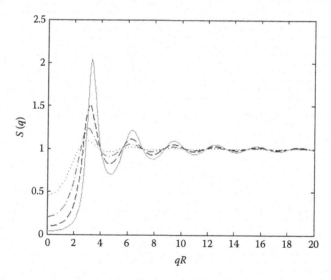

Figure 2.41 $S(q)$ plotted as a function of qR for some different concentrations of spherical particles interacting as hard spheres. Dotted line: $\phi = 0.1$; dashed dotted line: $\phi = 0.2$; dashed line: $\phi = 0.3$; and solid line: $\phi = 0.4$. The correlation peak moves to higher q with increasing concentration as the average separation between particles decreases.

volume interactions of the hard spheres become significant and the solution becomes increasingly structured, as seen by the oscillations in $S(q)$ and the decreased osmotic compressibility.

In Figures 2.42–2.44, we have plotted separately $P(q)$, $S(q)$, and $I(q)$ for a hard sphere system with volume fractions $\phi = 0.01$, 0.10, and 0.40, respectively. As can be seen, there is an increased influence of the structure factor on the scattering pattern as concentration increases. At the lowest concentration $\phi = 0.01$ (Figure 2.42), $S(q) \approx 1$ and does not influence the scattering significantly.

Up to now, we have only considered monodisperse particles, in which all particles exactly have the same size. This is seldom the case in reality. Most often there is a mixture of sizes, with a size distribution that, for example, can be Gaussian $N_p(R)/V \sim \exp\{(R - \langle R \rangle)^2/2\sigma^2\}$, where $N_p(R)/V$ is the concentration of particles with radius R, and the distribution is characterized by a mean value $\langle R \rangle$ and a standard deviation σ. The main effect

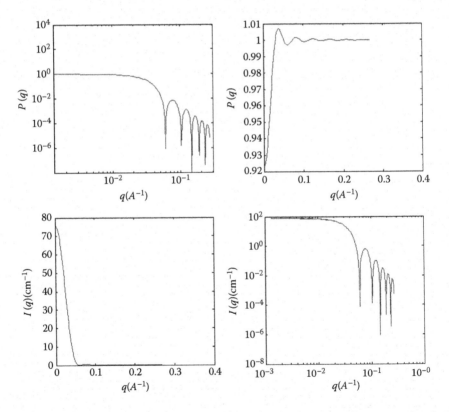

Figure 2.42 $P(q)$, $S(q)$, and $I(q)$ for a dispersion of spherical homogeneous hard sphere particles of $R = 10$ nm and $\phi = 0.01$.

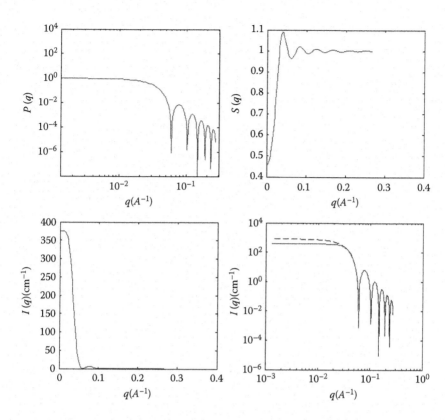

Figure 2.43 $P(q)$, $S(q)$, and $I(q)$ for a dispersion of spherical homogeneous hard sphere particles of $R = 10$ nm and $\phi = 0.10$.

of the polydispersity is that the formfactor minima in the scattering patterns become less distinct as the different particle sizes have their minima at different q values. In Figure 2.45, we compare $P(q)$ from monodisperse spheres with an average $P(q)$ from a system with a Gaussian distribution of sizes of a relative standard deviation $\sigma/\langle R \rangle = 0.1$. With this polydispersity, only the first two minima are clearly visible.

In Figure 2.46, we show the scattered intensity from a real experimental system. The data were recorded in SAXS experiments and the system is a microemulsion, where spherical droplets of decane covered by a stabilizing layer of the nonionic surfactant $C_{12}E_5$ (pentaethyleneglycol dodecyl ether) are solubilized in water. The mean radius of these microemulsion droplets is 8 nm and $\sigma/\langle R \rangle \approx 0.15$ (Balogh et al. 2006). The scattering from three different concentrations, $\phi = 0.011$, 0.11, and 0.22, respectively, are shown. The scattered intensity at higher q, where $S(q) \approx 1$ for all concentrations, is proportional to ϕ. At lower q, the structure factor reduces the

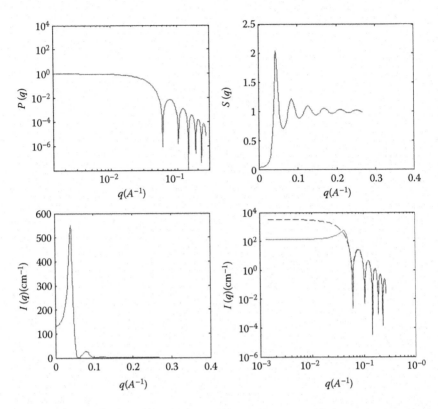

Figure ***2.44*** $P(q)$, $S(q)$, and $I(q)$ for a dispersion of spherical homogeneous hard sphere particles of $R = 10$ nm and $\phi = 0.40$.

scattered intensity for higher concentrations. Below, we will also discuss light scattering from the same microemulsion system.

In Equation 2.45, we also see the very strong size dependence of the scattered intensity. For a given volume fraction of particles, $\phi = N_p V_p / V$, the scattered intensity is proportional to the particle volume which, in the case of homogeneous spheres, means a proportionality to R^3. Air and water molecules scatter the sunlight only little and the main effect is the blue color of the sky. When water molecules in humid air condense to form droplets and ice particles, they scatter much more, and the resulting clouds are clearly visible.

Static light scattering

With visible light we have access to small q-values. Mostly, the wavelength of the light (ca. 500 nm) is longer than the size of the particles we are

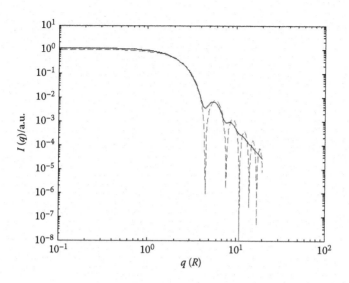

Figure 2.45 Comparison of $P(q)$ for monodisperse spheres with $R = 10$ nm (red dashed line) with the average $P(q)$ for polydisperse spheres with $\langle R \rangle = 10$ nm and relative standard deviation $\sigma/\langle R \rangle = 0.10$ (blue solid line). Polydispersity removes the formfactor minima in the scattering pattern.

studying. In this case, we will not see any formfactor minima within the accessible q-range (for a homogeneous sphere, the first formfactor minimum occurs at $q = 4.5/R$). Rather, we only see the low q part of the product $P(q)S(q)$. With light scattering experiments, we can make accurate extrapolations to $q = 0$ to determine properties like V_p (the molecular weight of polymers is often determined this way) and $(\partial \pi/\partial \phi)^{-1}$, but we are typically unable to determine the particle shape. Although we will not detect any formfactor minima, the monotonic variation (decay) of $I(q)$ with q can still be evaluated, to obtain the radius of gyration, R_g, from the leading term in the series expansion of $P(q)$:

$$P(q) = 1 - \frac{q^2 R_g^2}{3} + \cdots \qquad (2.47)$$

For a homogeneous sphere, $R_g = (3/5)^{1/2} R$.

In light scattering experiments, one often explores a wide range of scattering angles (15°–165°). A schematic description of a light scattering setup is shown in Figure 2.47. A laser beam is focused on the sample and the scattered light is collected, for example, by a photo diode placed on a movable arm (goniometer), to detect the scattering at different angles.

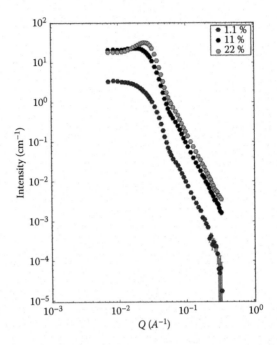

Figure 2.46 SAXS data from aqueous solutions of microemulsion oil droplets with radius $R \approx 8$ nm. The oil (decane) droplets are covered and stabilized by a layer of the nonionic surfactant $C_{12}E_5$ and interact to a good approximation as hard spheres. Three different concentrations are shown, $\phi = 0.011$, 0.11, and 0.22, respectively. (From Balogh, J., U. Olsson, and J.S. Pedersen. *Journal of Dispersion Science and Technology* 27(4), 2006: 497–510. With permission.)

The laser wavelength (e.g., 633 nm for a helium–neon laser) is generally expressed as the wavelength in vacuum. In the sample, this wavelength becomes λ/n, where n is the refractive index of the solution, and the expression for the q-vector is, for light scattering, generally written as

$$q = \frac{4\pi n}{\lambda} \sin \frac{\theta}{2} \qquad (2.48)$$

With a He–Ne laser, the accessible q-range becomes ca. 0.0035–0.026 nm^{-1} in water ($n = 1.33$). In light scattering, the absolute scale scattered intensity is generally referred to as the excess Rayleigh ratio, $\Delta R(q)$. The data are converted into absolute intensities using

$$\Delta R(q) = \frac{\Delta I(q)}{I_{ref}(q)} \left(\frac{n}{n_{ref}} \right)^2 R_{ref}(q) \qquad (2.49)$$

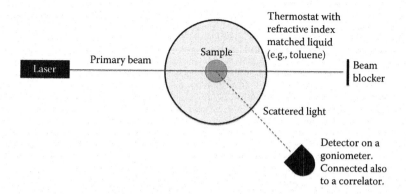

Figure 2.47 Schematic outline of a typical goniometer light scattering instrument. A laser beam of 1–2 mm width shines on a sample inserted into a liquid (often toluene or decaline) having a refractive index very close to that of the sample container (glass) to minimize reflections, etc. The index matched liquid is also used as a thermostat. The scattered light is recoded by a detector (e.g., photodiode), fixed on a goniometer arm. For dynamic light scattering experiments, intensity correlations are analyzed in a correlator.

Here, $\Delta I_{ref}(q)$ is the excess scattered intensity of the sample, where the scattering from the solvent and sample tube has been subtracted, and n its refractive index. $I_{ref}(q)$ is the scattered intensity of the reference solvent and n_{ref} its refractive index. R_{ref} is the Rayleigh ratio of the reference solvent. A common solvent for absolute scale calibration of the scattered intensity is toluene.

The contrast arises from the difference in the refractive index between the particle and the solvent. In light scattering, this is quantified in terms of the so-called refractive index increment, $dn/d\phi$ (more commonly expressed as dn/dc, where c is the molar concentration).

The light scattering power of molecules depends on their polarizability, which is linked to the refractive index. The scattered light intensity has a very strong wavelength dependence, $I \sim \lambda^{-4}$. This is the reason why the sky is blue (on a sunny day). When we look at the sky, we detect photons from the white light of the sun that have been scattered by air molecules in the atmosphere. Blue photons ($\lambda = 400$ nm) are scattered more frequently than other visible wavelengths and that is why the sky is blue. This is also why the sun looks yellow. Blue photons have preferentially been depleted from the originally white light and the result is that the transmitted sunlight is yellow. At sunset, the sun light takes a longer path through the atmosphere. As a consequence, there is more scattering. Often, we then find that the sun looks beautifully red. This is because so much of the other colors have been scattered, that only the longest wavelength

(800 nm) has a significant transmission. This is helped if the atmosphere is polluted also by soot or other particles.

The excess Rayleigh ratio can be expressed as

$$\Delta R(q) = \phi \frac{4\pi^2 n_0^2}{\lambda^4} \left(\frac{dn}{d\phi} \right)^2 V_p P(q) S(q),$$ (2.50)

where n_0 and n are the refractive indices of the solvent and the solution, respectively. Extrapolating data to $q = 0$, where $P(q) = 1$, we can, determine V_p if the solution is dilute so that $S(q) = 1$. As mentioned above, this is commonly done to determine the molecular weight of polymers. If V_p, on the other hand, is known from other measurements, $\Delta R(0)$ determined as a function of ϕ reports on interparticle interactions via the concentration dependence of $S(0)$. Figure 2.48 shows the variation of $\Delta R(0)$ with ϕ in a microemulsion composed of spherical oil droplets ($R = 8$ nm) in water, covered and stabilized by a layer of nonionic surfactant (Olsson and Schurtenberger, 1993). The system is in fact the same as in the SAXS study discussed above (see Figure 2.46). The solid line is the prediction for hard spheres, given by the Carnahan–Starling (1969) equation. At lower concentrations, the intensity increases with concentration because of the increased number of particles. At higher concentrations, however, the scattered intensity decreases because of the repulsive interactions. This has a very important consequence. Our

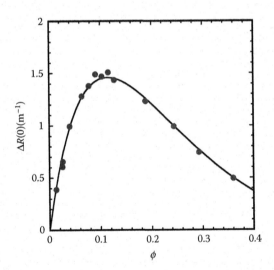

Figure 2.48 The excess Rayleigh ratio extrapolated to $q = 0$ plotted as a function of the volume fraction for the same water–$C_{12}E_5$–decane microemulsion as in Figure 2.46. The solid line is the theoretical prediction for hard spheres. (From Olsson, U. and P. Schurtenberger. *Langmuir* 9(12), 1993: 3389–94. With permission.)

eye lenses are composed of concentrated ($\phi \approx 0.3$) protein solutions. For obvious reasons) they need to be transparent, with only very low scattering of light. This is secured by the strong, here electrostatic, repulsions between the protein molecules. In the cataract disease, the eye lens has turned turbid with significant light scattering. This is because some of the protein interactions have shifted from net repulsive to attractive and the proteins have aggregated into larger objects (Stradner et al., 2007).

Dynamic light scattering

The scattered intensity measured in static light scattering experiments corresponds to a time average, $\langle I(q) \rangle$. The fluctuations of the intensity arise because the relative positions of all the particles are constantly changing as they undergo Brownian motion. The static structure factor, discussed above, is therefore strictly a time average. Fluctuations in the intensity may also arise from particle shape fluctuations, that is, fluctuations in $P(q)$, but this can often be neglected, and will not be discussed further here. By analyzing the rate of intensity fluctuations, we get information about the rate of particle motions. In a dynamic light scattering experiment, the autocorrelation function of the intensity fluctuations is recorded. This is done recording the intensity at different times, multiply those values with the value at $t = 0$, repeat these operations many times, and finally take the statistical average of the products to form the autocorrelation function, $G(t) = \langle I(0)I(t) \rangle$. Here, $\langle \rangle$ symbolizes that it is an average. This correlation function has the value $\langle I^2 \rangle$ at $t = 0$ and decays to $\langle I \rangle^2$ at long times when $I(0)$ and $I(t)$ has become uncorrelated. Often one considers the modified correlation function $g_2(t) - 1$, where $g_2(t) = G(t)/\langle I \rangle^2$. For the case of monodisperse particles, this is an exponential:

$$g_2(t) - 1 = A e^{-2\Gamma t} \qquad (2.51)$$

where A is an instrumental constant and the relaxation rate Γ is given by

$$\Gamma = D_c q^2 \qquad (2.52)$$

The q^2 dependence is indicative of diffusive motion and D_c is the collective diffusion coefficient. D_c depends not only on the particle size, but also on interactions, direct as well as hydrodynamic. In dilute solutions where interactions can be neglected, we obtain the hydrodynamic radius, R_H, from D_c that here equals the Stokes–Einstein diffusion coefficient:

$$D_0 = \frac{k_B T}{6\pi\eta R_H} \qquad (2.53)$$

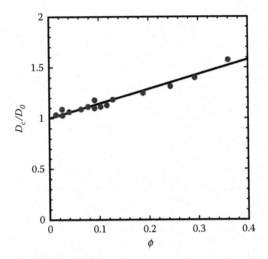

Figure 2.49 The relative collective diffusion coefficient, D_c/D_0, plotted as a function of the volume fraction for the same water–$C_{12}E_5$–decane microemulsion as in Figures 2.46 and 2.48. $D_0 = 2.0 \times 10^{-11}$ m^2 s^{-1}. The solid line is the theoretical prediction for hard spheres. (From Olsson, U. and P. Schurtenberger. *Langmuir* 9(12), 1993: 3389–94. With permission)

Here, k_B is Boltzmann's constant, T the absolute temperature, and η the solvent viscosity. For hard sphere systems, D_0 has only a weak concentration dependence. In Figure 2.49, we show the variation of D_c with concentration in the same nonionic microemulsion system as in Figures 2.46 and 2.48 (Olsson and Schurtenberger, 1993). The solid line is a theoretical prediction (Pusey, 1990) for hard spheres, $D_c/D_0 = 1 + 1.45\phi$. From the value of D_c extrapolated to $\phi = 0$ ($D_0 = 2.0 \times 10^{-11}$ m^2 s^{-1}), a hydrodynamic radius of ca. 90 nm is obtained, using Equation 2.53.

References

Balogh, J., U. Olsson, and J.S. Pedersen. Dependence on oil chain length of the curvature elastic properties of nonionic surfactant films: Emulsification failure and phase equilibria. *Journal of Dispersion Science and Technology* 27(4), 2006: 497–510.

Carnahan, N.F. and K.E. Starling. Equation of state for nonattracting rigid spheres. *The Journal of Chemical Physics* 51(2), 1969: 635–36.

Kinning, D.J. and E.L. Thomas. Hard-sphere interactions between spherical domains in Diblock copolymers. *Macromolecules* 17(9) 1984: 1712–18.

Narayanan, T. High brilliance small-angle X-ray scattering applied to soft matter. *Current Opinion in Colloid & Interface Science* 14(6), 2009: 409–15.

Olsson, U. and P. Schurtenberger. Structure, interactions, and diffusion in a ternary nonionic microemulsion near emulsification failure. *Langmuir* 9(12), 1993: 3389–94.

Pedersen, J.S. Analysis of small-angle scattering data from colloids and polymer solutions: Modeling and least-squares fitting. *Advances in Colloid and Interface Science* 70, 1997: 171–210.

Pusey, P.N. Liquids, freezing and the glass transition. In *Les Houches Session Li.* Edited by D. Levesque, J. Hansen, and J. Zinn-Justin. Amsterdam, The Netherlands: Elsevier, 1990.

Stradner, A., G. Foffi, N. Dorsaz, G. Thurston, and P. Schurtenberger. New insight into cataract formation: Enhanced stability through mutual Attraction. *Physical Review Letters* 99(19), 2007: 198103.

Microscopy

Cem Levent Altan and Nico A.J.M. Sommerdijk

Introduction

Seeing an object provides immense information about that object. The size, shape, color, texture, and physical state become immediately apparent once an object is visually inspected and other information can then be inferred from this. Science concerns itself with numerous objects of a wide range of sizes, most of which are "invisible" to the naked eye. In a quest to visualize "invisible" objects such as planets that are light years away from earth, or materials in atomic scale, technological advancements facilitated visual gathering of information with the invention of telescopes and microscopes.

Size is defined as the relative extent or dimensions of an object. Length, which is one of the dimensions of an object, is defined with the standard unit of a meter. It is well within our ability to associate a certain distance with a meter and distances with three orders of magnitude more or less than a meter. However, when the whole spectrum of existence is considered, this range is only a small portion of a huge scale of lengths, starting from the smallest atom to the universe. Figure 2.50 shows the length scale with corresponding examples to each range in the logarithmic scale. In this wide spectrum of lengths, the upper unknown border is explored via telescopes, while the bottom line exploits microscopes. Materials that belong to this sub-portion of the length scale are said to be in microscopic scale.

Light is an electromagnetic radiation (EMR) that can be used as a probe to understand the shape and morphology of objects. Light can be described by properties of waves; however, when light interacts with a material, electrons get emitted (photoelectric effect), implying particle-like property. This points to the dual nature of light, both wave and particle character. When photons interact with objects, they are reflected, diffracted, or transmitted and these photons are detected by the eye. In order to differentiate between two different points, the probe that is used should have dimensions in the same order as the object that is analyzed. When light is used

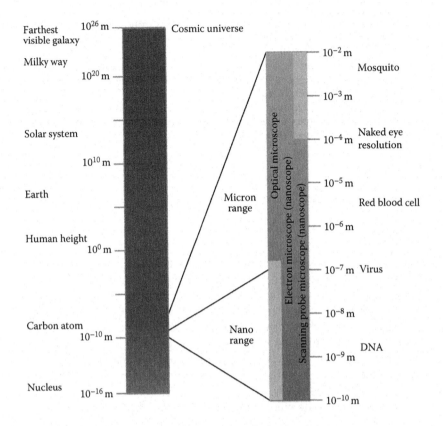

Figure 2.50 Length scale from atoms to galaxies.

as a probe, only objects that are larger than half the wavelength (reoccurring period of the wave) of light (i.e., ~500 nm) can be visualized. Although light is an efficient probe for imaging objects in the macroscopic scale, molecules and atoms require short-wavelength probes for detailed visualization. Resolution can be defined as the ability to distinguish small sections of a specimen that are in close proximity with each other as two distinctly separate parts. Resolution is limited by the wavelength of the probe.

Magnification is how much an image is enlarged under the microscope. Magnifying an image does not enhance its resolution and the magnification limit of a microscope is determined by its lenses. In novel characterization techniques, resolution and magnification limitations are challenged by the use of particles with shorter wavelengths as probes and introduction of complex lenses.

Looking at (Greek: skopeo) small objects (Greek: mikros) has always intrigued people. The Gallilean telescope can be considered the first microscope, dating back to the beginning of the seventeenth century. In the second half of that century, Antoni van Leeuwenhoek built an instrument comprising a convex lens and a sample holder achieving 400× magnification, which allowed the inspection of protozoa and bacteria. Assembly of an objective lens, an eyepiece, and a light source constituted the basis of the light microscopes that are used today.

Based on their working principles, microscopes can be classified into three different groups. The light microscope, also known as optical microscope, is the most common, simplest and first microscope, which uses visible light as a probe and an assembly of lenses to focus and magnify the image. A light microscope can resolve objects as small as 0.2 μm and magnify up to 2000 times. Light microscopes do not need complicated sample preparation and can be used for color imaging of dynamic systems for both living (e.g., cells) and non-living (e.g., liquid crystals) specimens. Although this resolution limit is sufficient for living cells, large viruses, or small bacteria, it is not possible to visualize an atom that is 1000 times smaller.

To probe structures in the atomic scale, shorter-wavelength probes are achieved by accelerating electrons. The velocity of an electron and its wavelength are inversely proportional and accelerators decrease the wavelength of the electron to 100,000 times less than that of light. Microscopes that use electron beams rather than light beams are called electron microscopes (EMs) and they have much better resolution than light microscopes. These microscopes also use magnetic coils instead of glass lenses to provide higher magnification. Better resolution and high magnification allows for the inspection of relatively smaller specimens such as viruses, proteins, DNA, and atoms having sizes in the order of nanometers. It is also possible to have information about the 3D structure of the specimen by using EMs. Although EMs are important tools for the nanoworld, the need of working under vacuum and extensive sample preparation are some of the drawbacks of the technique.

Instead of light or electrons beam, scanning probe microscopes (SPMs) use a small, sharp scanning needle, which moves over the object at a distance of near contact with the surface as a probe. The interactions between the needle and the atoms over the object's surface are interpreted to form a three-dimensional image of the sample surface while optical and EMs are suitable for x–y lateral imaging, SPMs are best suited for z-height measurements. Specimens can be analyzed at ambient conditions even in the liquid phase, which is not possible with EMs without using special liquid cell holders (de Jonge and Ross, 2011). SPMs can also analyze other physical properties such as thermal and magnetic, which is not directly possible with other microscopes. Because EMs and SPMs image in the nanometer range, sometimes these microscopes are referred to as nanoscopes. Although SPMs provide excellent information in the z-direction, lateral imaging is unparalleled in EMs, which make these two techniques complementary in investigation of structures in the atomic scale.

The image of a sample is directly related to brightness and contrast as well as resolution and magnification. Brightness is the amount of signal that reaches the detector during imaging which is related to the illumination system and can be changed by altering the beam intensity. Focusing the beam through an aperture, which is a disc containing a precise circular hole through which electron or light beam travels, increases its intensity. Brightness also depends on the atomic number of the specimen in EM and topography of the area in SPM.

On the other hand, contrast is the difference in brightness between the details of the specimen or the specimen and its background. When the contrast is higher, the difference between the signal intensity of two different parts is higher. Details are related to the resolution that is dependent, for example, on the wavelength of the illumination source in optical microscopy, on the accelerating voltage for EMs, and on the tip sharpness for SPMs. High voltages are employed in EMs for better resolution at the cost of contrast. Contrast can be enhanced by using appropriate objective apertures that show the importance of optimization in imaging. The differences between brightness, resolution, and contrast can be seen in Figure 2.51.

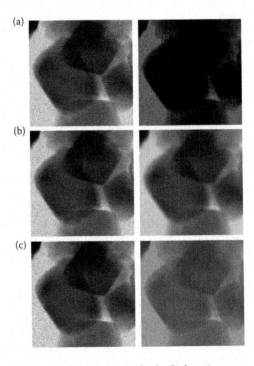

Figure 2.51 Magnetic iron oxide nanoparticles by light microscope at the same magnification: (a) brightness (good–left), (b) resolution, and (c) contrast. (Figure courtesy: Nico A.J.M. Sommerdijk, Techische Universteit Eindhoven, Eindhoven, Netherlands).

Optical (light) microscopy

Optical microscopes (light microscopes) were the first microscopes used to obtain a magnified image of an object. Optical microscopes use visible light (EMR) as a probe (Figure 2.52). Light is focused by a condenser lens and reaches the specimen on the microscope stage. Objective lens collects the light reflected or transmitted by the specimen and the final magnified image of the object is projected directly to the retina of the eye or to an external imaging device such as a screen or camera (Murphy and Davidson, 2012). If the object is analyzed by human eye, then the final image formation is done by an eyepiece instead of a projector lens. With the help of the eye's cornea and lens, the second real image is formed onto the retina where it is processed by the brain as a virtual magnified image. Generally, novel microscopes include both eyepiece and projector lens for both purposes. Schematic representation of the components of an optical microscope and its corresponding arrangement is given in Figure 2.52b. Figure 2.53 presents the magnification in a microscopy.

The magnification limit of a light microscope depends on the combination of the ocular lens and the objective lens together. The higher limit is defined as the multiplication of the highest possible magnification that those two different lenses can offer. Because ocular (eyepiece) lenses mainly have a standard magnification value of 10×, the total magnification is essentially determined by objective lens, which makes it the most important part of an optical microscope.

Light or optical microscopes are suitable for imaging both solid and liquids. The analysis of the microstructure with an optical microscope requires appropriate sample preparation. Solid specimens that are sufficiently thin can directly be placed and pressed in between microscope slides. However,

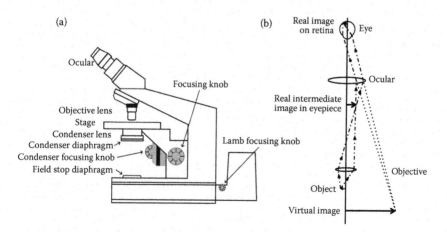

Figure 2.52 Light microscope: (a) components and (b) working principle.

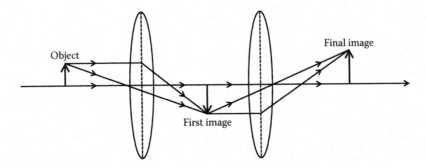

Figure 2.53 Magnification in microscopy.

thick samples should be divided into thinner portions by cutting over a cross-section, a method known as microtomy. After the specimen is cut to reach the appropriate thickness, it is embedded on a mounting material. In order to remove any damage caused by previous sectioning, these samples are also polished. On the other hand, liquid dispersions can be placed directly between a microscope slide and a cover slide for imaging.

There are different techniques for imaging a specimen using a light microscope. Bright-field microscopy is the most frequently used technique. In bright-field microscopy, the view area is illuminated with high-intensity light and the final image is composed of both diffracted (interacted with specimen) and non-diffracted (passed through the specimen) light rays. The differences in density between the object and its background result in the object in question to be highly visible. However, specimen that do not effectively impede with light would need a further treatment called staining. Staining increases the contrast by selectively highlighting the whole or a part of the specimen by specific interactions of an agent with the specimen. For instance, when special dyes or pigments are used that solely interact with the cell wall, the cell becomes highly visible. There are several types of stains used to highlight different components of a cell, depending on the emphasis needed. Figure 2.54 shows stained malarial cells under 630× magnification.

However, as dyes or pigments may influence the dynamics of living specimen, another method called phase-contrast imaging is often used rather than staining, which exploits the refractive index difference of the specimen and its surroundings. The light that passes through a transparent specimen has a different phase than the light that passes through the background. This change in phase is brought to distinguishable values by the help of a phase plate by shifting its wavelength, causing a difference in brightness thus enhancing contrast. Alternatively, in dark-field microscopy the view area is kept dark by a stopper in the condenser of the microscope and the specimen is illuminated by a particular condenser, which

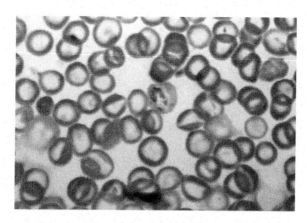

Figure 2.54 Giemsa-stained micrograph depicts an example of a slightly acidic slide that yielded a pink colored resultant stain. (Photograph courtesy: Public Health Images).

causes light to be diffracted from the specimen at an angle (Figure 2.55). The non-diffracted light rays (directly transmitted light) are removed by objective apertures and only light that is diffracted by the specimen is collected to compose the final image, resulting in a high contrast image (Murphy and Davidson, 2012). In bright-field microscopy, the inherent light parts of the specimen are invisible and in dark-field microscopy, the naturally dark parts become invisible, and therefore, these two techniques can be used in conjunction to achieve a more informative image of the specimen. Images obtained from bright-field, dark-field, and phase-contrast microscopy of the same sample are given in Figure 2.56.

Finally, in fluorescence microscopy, the specimen in part is fluorescent or bound to fluorescent complex agents, which are used to illuminate the corresponding area after being excited by a light of high energy and short wavelength.

Light microscopes are significantly cheaper than other imaging instruments and are designed to visualize the specimen in full color. Despite having the advantage of imaging dynamic properties of living (cells) and non-living (liquid crystals) specimens, specimens in nanometer sizes that require high-resolution imaging at elevated magnifications cannot be investigated using light microscopes.

Scanning probe microscopy

SPM is a method to analyze materials via interactions between a physical solid probe and the surface by scanning. The surface image is formed by the mechanical movement of the probe over the surface of the specimen followed by the interpretation of the interactions of the probe and the

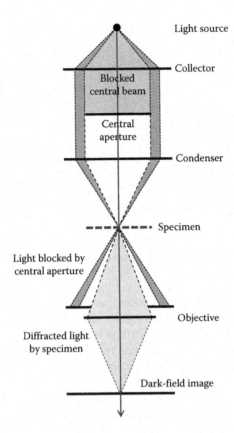

Figure 2.55 Light path for dark-field microscopy.

surface as a function of vertical and lateral positions (Figure 2.57). It has the power of examining properties of the surface down to atomic spacing. The resolution is determined by the spatial dimensions of the probe and the distance between the probe and the surface. The lateral and vertical

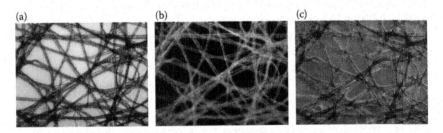

Figure 2.56 Different imaging techniques used to generate contrast: (a) bright field (b) dark field, (c) phase contrast illumination. (Wheeler R., http://www.richardwheeler.net, photograph courtesy: wikipedia).

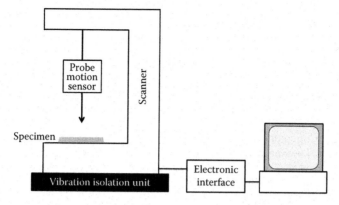

Figure 2.57 Essential components of an SPM.

resolutions of an SPM can be down to 0.1 and 0.01 nm, respectively. The advantages of most forms of SPMs are that they require minimal or no sample preparation, can be operated at ambient conditions, and is suitable for liquid samples (Leng, 2009).

The general SPM system given in Figure 2.57 consists of a probe, a motion sensor, a scanner, an electric controller, and a computer. The probe is the most important component of an SPM as it interacts with the surface of the sample. For the analysis, either the probe scans over a fixed surface or the probe is fixed and the sample surface is moved by the stage (Leng, 2009). The probe geometry affects the lateral resolution due to the tip con- volution effect. The signal that is received due to interaction is a convolu- tion of sample surface topography and tip topography. Thinner probes are more capable of capturing the details of the topography of the sample. For a good resolution, the tip should be as thin as 10 nm. As shown in Figure 2.58, the broader probe tips cause features to appear broader while the holes appear shallower (Eaton and West, 2010).

The scanner of the SPM holds the probe and positions it in three dimensions and the movement is obtained by using three piezoelectric materials in an orthogonal arrangement. As piezoelectric tubes are able to change their shape in an external electric field, the field formed by the high-voltage amplifier causes the piezoelectric tubes to be reshaped, thus move the probe. The supplied voltage to the amplifier is controlled by a computer, which is also used to receive the position signal of the probe.

Scanning tunneling microscopy (STM) was the first example of the SPM invented in 1981, which uses a current that flows through a gap between a metallic tip and the surface atoms of a conducting material such as metals. The tunneling current is interpreted to determine the spacing between the probe and the sample. As STM is limited to the inves- tigation of conducting specimens, a new technique where near-field forces

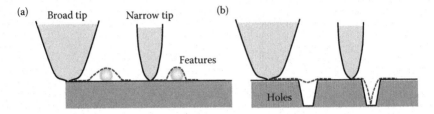

Figure 2.58 Convolution effect of broad and narrow SPM probe on (a) features and (b) holes. The image construction is shown by a dashed line.

between the atoms of the tip and the surface are used rather than an electrical current was developed in 1986. Atomic force microscopy (AFM) is now more widely used as it can be applied to more versatile surfaces. The material of the probe and its working principle in STM and AFM are also different. The STM probe is made of a tungsten wire and conducts the current that flows between the surface and the probe tip atoms. In AFM, the probe is generally made of SiN that is placed on a cantilever spring. Interactions between the surface and the tip result in a force, which deflects the laser beam shining on the probe. This deflected beam is then collected and used to define spacing (Leng, 2009). SPMs can also be used to analyze physical properties such as magnetic properties of the sample (magnetic force microscopy) which is not possible with optical and EMs. As the property of the sample surface defines the interaction between the surface atoms and the scanning tip, the correct type of SPM for analysis should be carefully chosen.

Scanning tunneling microscopy

STM is based on a tunneling current between a conducting scanning tip and a conducting sample. The tip is located at a few angstroms away from the sample (Figure 2.59). Tunneling current is the electrical current that flows as a result of the voltage difference between two conductors that are separated by a very short (sub) nanometer distance where an insulator, such as vacuum, exists between the conductors (van de Leemput and van Kempen, 1992). The intensity of the current is directly related to the distance between the conductors and the thickness of the insulator. This dependence is used to interpret the surface properties of the sample. STM has the ability to position the scanning tip over an individual atom and therefore has the resolution of atomic scale for surface interpretation. For an analysis, the scanning tip moves along the sample in the x–y-direction and a tunneling current flows in the z-direction between the tip and the surface. Overall topography of the sample surface is then formed by collecting and assembling these series. Scanning can provide information on the atomic surface structure composed of an array of all individual atoms forming the sample. In STM, contrast is obtained by differences in the

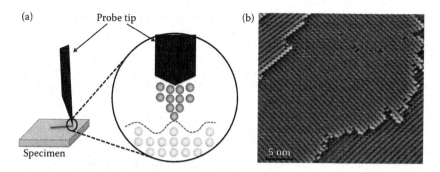

Figure 2.59 (a) Illustration of scanning tunneling microscopy and (b) STM topography of a clean Si (001) surface. (From Dürr, M. and U. Höfer. *Progress in Surface Science* 88(1), 2013: 61–101. With permission.)

current throughout the sample where light regions show the areas where high current has flown and vice versa (Frommer, 1992).

There are two main modes of operation in STM: the constant-current and the constant-height modes (Figure 2.60). There is a feedback loop in the constant-current mode, which is used to change the position of the probe in vertical direction to obtain a constant tunneling current. When the feedback controller loop is disregarded, the constant-height mode is obtained where the position of the probe tip is kept constant. Constant-height mode scans the surface at a faster rate which makes it particularly useful for analyzing dynamic samples as height change in constant-current mode needs more data collection time. However, in this mode, there is the possibility of damaging the sample tip if there is an increase in the sample height as the tip position is fixed (Leng, 2009).

STM does not collect data by averaging the input taken from the sample. It is focused on all individual atoms, thus has the ability to show the atomic defects and grain boundaries similar to TEM, which is not possible

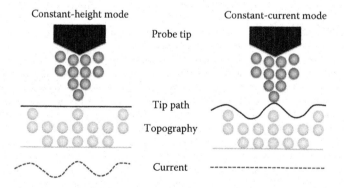

Figure 2.60 Operating modes of STM.

by other techniques that are used for atomic structure analysis (e.g., X-ray diffraction) (Frommer, 1992). However, STM is effective only on conductive or semiconductive materials and relies on the electrical properties of the sample surface. Additionally, surface analysis is highly affected by external effects such as any kind of vibrations or movements in the environment (Wang, 2006).

Atomic force microscopy

The AFM is another type of SPM, which provides information on the surface properties at the nanometer scale. All SPMs measure a physical property that arises from interactions (e.g., tunneling current for STM) and AFM measures the force between a sharp tip and the surface of the sample (Vilalta-Clemente and Gloystein, 2008). This sharp scanning probe, which is mounted over a flexible cantilever, moves along the sample at short distances in the range 0.2–10 nm. When the sharp tip and the surface are sufficiently close, an interaction due to intermolecular forces occur. This interaction force deflects the cantilever and the corresponding deflection is measured by a detector and laser. The force is solely dependent on the cantilever constant and the distance between the sample and the scanning probe tip. As illustrated in Figure 2.61, the bending of the probe and vertical–horizontal deflection of the cantilever monitored during the surface analysis and then detected by a laser that is reflected from the back of the scanning probe to obtain the topography of the sample surface.

The interaction of the sample and the scanning probe is determined by Van der Waals interactions. When the tip is close enough to the surface, repulsive Van der Waals forces occur, whereas attractive forces dominate

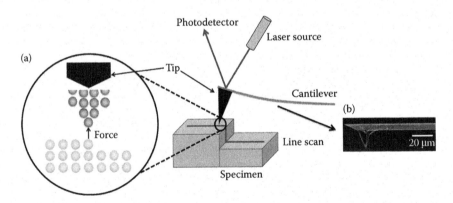

Figure 2.61 (a) Illustration of atomic force microscopy and (b) SEM image of a blank AFM tip. (From Matsuura, H. et al. *Surface Science* 583(1), 2005: 29–35. With permission.)

when far away. AFM can be operated at three different modes for imaging (Wilson and Bullen, 2006) (Table 2.4). In the contact mode, the tip is in soft direct contact with the surface. As the sample is scanned, the contact force bends the cantilever responding to the changes in the topography of the sample. These deflections can be used to form topographic visualization of the sample surface by two different modes. In the constant-force mode, the force between the scanning tip and the surface, thus the deflections of the cantilever, is kept constant using a feedback loop. The feedback loop collects the deflection of the cantilever as an input and moves the scanner in the z-direction to keep the deflection constant. For the generation of the image, the z-value of the scanner is collected. On the other hand, the z-value of the cantilever is kept constant in constant-height mode. This mode provides fast scanning but relatively high interactions between the

Table 2.4 Comparison of AFM Modes

	Advantage	Disadvantage
Contact mode	• High scan speeds • Samples with extreme changes in vertical topography can be more accurately scanned	• Lateral forces may distort features in the image • In ambient conditions may get strong capillary forces due to adsorbed fluid layer • Combination of lateral and normal forces reduces resolution • The tip may damage the sample and the sample may damage the tip • Slower scan speed than in contact mode
Tapping mode	• Lateral forces almost eliminated due to tapping • Higher lateral resolution on most samples • Less applied force so less damage to soft samples or tips	
Non-contact mode	• Both normal and lateral forces are minimized, so good for measurement of very soft samples • Can get atomic resolution in a UHV environment	• In ambient conditions the adsorbed fluid layer may hinder the measurements • Slower scan speed than tapping and contact modes to avoid contacting the adsorbed fluid layer

Source: Li, A., P. Coombe, and R. Oliver. Techniques for Studying Materials: Atomic Force Microscopy. 2010. University of Cambridge, http://core.materials.ac.uk/repository/doitpoms/tlp/afm.htm

tip and the surface, which may damage soft samples. In intermittent (tapping) mode, the cantilever freely oscillates and taps the surface. When the oscillating cantilever reaches and contacts the surface, the energy loss due to the interaction between the tip and the sample surface causes reduction of oscillation. The decrease in the amplitude of the oscillation is then used to interpret the surface features of the sample. This mode provides a high resolution for soft samples especially biological specimens but needs more scanning time. Finally, in the noncontact mode, the tip and the surface does not have a contact. The probe measures the oscillations in the attractive Van der Waals and other long-range forces between the sample and the probe using a feedback loop. Although for very sensitive samples, non-contact mode may be employed, these weak long-range forces result in lower resolution. In addition, the adsorbed fluid layer on all samples may hinder this measurement, and therefore this mode is most effective in ultra-high vacuum conditions.

AFM can also be used to analyze the mechanical properties such as hardness, elasticity, and adhesion according to the force that is measured (Wilson and Bullen, 2006). Soft samples are preferably fixed on a sample container using adhesive tape. On the other hand, hard samples can be directly mounted and scanned. Powder samples can be dispersed in a liquid that does not solubilize it and a drop is placed on a mica or silicon substrate. For powders that contain large particle size or not able to be dispersed in a liquid, tablet press method is used.

AFM tips are mainly silicon, Si, or silicon nitride, Si_3N_4. Depending on the application, these tips may be coated with gold or chromium. Low-aspect-ratio tips are used for general testing, which are not suitable for the best surface topography imaging. High-aspect-ratio tips made of carbon nanotubes or tungsten provide a better resolution, but quite expensive.

One of the most significant advantages of the AFM is that biological specimens, which require dynamical inspection at normal conditions, can be analyzed (Li, Coombe et al., 2010). The scan speed is the limitation of this technique and in order to increase the scan rate, parallel probes are sometimes employed.

Electron microscopy

Ordinary light microscopes, which use light as a probe and glass lenses to magnify, are unsuitable for analyzing specimens that require better resolution and higher magnification. In the 1920s, it was found that electrons moving in vacuum have wave-like properties similar to light. When magnetic and electric field are applied, the path of electron waves can be manipulated and focused like glass lenses, which led to electromagnetic lenses with higher magnification. Also, electrons can act as probes with much shorter wavelengths than light, reaching much higher resolutions.

Ernst Ruska (Berlin University of Technology) focused on these principles and merged these two properties which lead to the invention of the first EM in 1931 for which he received the Nobel Prize in 1986 (FEI, 2010).

Electrons that are accelerated by applying a voltage have significantly shorter wavelengths. For instance, electrons that are accelerated at 100 kV have 0.037 Å wavelength. These achievable shorter wavelength values resulted in enhanced resolution in comparison to optical microscopes. Although the wavelength of an electron is roughly 100,000 times smaller than visible light, the resolution of EMs is only in the order of 1000 times greater due to deviations from theoretical optics in lenses known as lens aberration. After the interaction of electrons with the sample, there are several possibilities. Some of the electrons are absorbed by the sample causing a mass thickness contrast. This absorption is directly related to the sample thickness and the nature of the sample. Some electrons are scattered by the sample in small angles that depend on the structural properties (e.g., composition) of the sample and cause a phase contrast. This scattering is in specific directions if the sample is crystalline and depends on the crystal structure and causes diffraction contrast. Images of the samples are a result of electron scattering from the atoms of the sample in EMs. Electrons can also be transmitted through the sample, and the amount of energy lost during the interactions of the electrons with the sample can be used for additional analysis of the samples, such as energy-dispersive X-ray spectroscopy (EDX) and electron energy loss spectroscopy (EELS). EDX and EELS are complementary to each other. EDX and EELS analyses show the atomic composition of a material, while EELS is further capable of getting information about chemical bonding, electrical, and surface properties. The scanning electron microscope (SEM) and transmission electron microscope (TEM) are the examples of EMs. In TEM, an electron beam is transmitted through a thin specimen of at most a few hundred nanometers, while an electron beam scans over a complete or selected region of a sample in SEM.

Although the capabilities of optical microscopes and EMs are different, the working principles are quite similar. As stated in Section "Optical (Light) Microscopy," optical microscopes use visible light and lenses to form a magnified image of the sample. EMs use electrons and electromagnetic lenses for the same purposes (Fahlman, 2011). The basic components of an SEM and a TEM are depicted in Figure 2.62. Probably the most important part of an EM is the electron gun. The gun itself is made of a filament and an anode. When a potential difference is applied between the anode and the filament, electrons are accelerated from the filament to the anode from where the formed electron beam is directed to the electromagnetic lenses in the microscope column (FEI, 2010).

There are three different lenses in EMs. Both instruments contain condenser lenses to control the intensity of illumination and objective

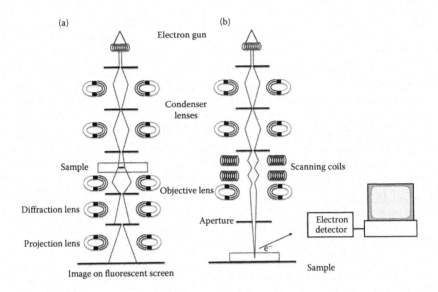

Figure 2.62 Components of (a) transmission and (b) scanning electron microscopy.

lenses to magnify the sample. They also hold additional projector lenses in order to project the image onto a screen or a camera (Fahlman, 2011).

As stated before, within EMs the moving electrons are manipulated by electromagnetic lenses and apertures (Leng, 2009). This electron movement should be in a vacuum for controlled focusing of the beam. Additionally, when not in vacuum, electrons scatter by collisions, which make the vacuum system crucial for good-quality imaging in EMs (Dunlap and Adaskaveg, 1997).

One of the most important features of TEM is its resolution, which is affected by the voltage applied to the electrons, determining the wavelength of the probe. In order to provide very stable voltage and current, there should not be any deviations and this can only be provided by complicated electronic circuits.

Scanning electron microscopy

In an SEM, electrons produced from a gun inside the microscope form a focused beam, which scans the surface of the object. Owing to its large depth of field, the distance between the nearest and farthest objects in focus related to the beam divergence angle, SEM is capable of producing images with 3D information which allows better analysis of the surface morphology and structural change of specimens. The reason for this great depth of field arises from the geometry of the beam optics.

In SEM, the electron beam is condensed to a narrow probe for scanning the surface of the object. Generally, the voltage that is used for generating the electron beam is between 1 and 40 kV. SEM is composed of several condenser lenses and one final objective lens. The objective lens acts as a condenser lens and demagnifies the electron beam in order to be used as a probe in the nanoscale. This probe moves over the surface along a line and after scanning moves toward another line. Electrons emitted from the object reaches the detector and the image is formed on a screen. This is where the magnification limit of an SEM is determined. It is the ratio of the area that is scanned to the ratio of the area of the screen (Leng, 2009). Magnification on the SEM depends only on the excitation of the scan coils.

The beam of electrons, produced by filament and adjusted by the lenses, first reaches the specimen and then moves inside the specimen up to some distance. When electrons reach an electron or a nucleus of an atom inside the object, their path changes, which is called scattering. The electrons from the primary beam that are scattered directly back from the object rather than traveling inside are called backscattered electrons. These electrons have high energies as they travel a short path. Another interaction which is a consequence of primary electron beam and the specimen is the formation of secondary electrons. Secondary electrons are the ones that belong to and are displaced from the specimen as a result of primary electrons. Those electrons have relatively low energy thus can only be formed close to the surface. For topographical information, mainly secondary electron imaging mode is used. In this mode, specimen is illuminated well and interpretation is easier. Like secondary electrons, particular electrons displaced from a distinct orbit of an atom in the specimen, form X-rays. As these X-rays contain essential information about the atom that they emitted, intrinsic information about the specimen such as atomic number and compositional distribution can be gathered by analyzing them (Dunlap and Adaskaveg, 1997). Energy-dispersive spectroscopy (EDS) is used to analyze the corresponding elements according to the atomic number and their relative proportions. Figure 2.63 shows X-ray scattering produced in a scanned area of Co_2SiMn glass-coated microwires. The spectrum includes the number of X-rays that were collected by the detector on the y-axis, while the energy level is given in the x-axis, which is used to characterize the elemental aspects of the specimen. EDS can also be applied on different regions of the specimen for detailed characterization (Hafner, 2010).

Specific specimens, especially if a luminescent molecule is attached, can fluorescence when they are under the influence of an electron beam. This process is called cathode luminescence. Light photons that arise due to the luminescence of specimen or its attachment are collected by a detector that is additionally attached to SEM (Dunlap and Adaskaveg, 1997).

Image formation in SEM is similar to photography where normal visualization is formed by reflected light from the objects. As reflected electrons are used in SEM, the objects in an SEM image are brighter while the background is darker (Figure 2.64).

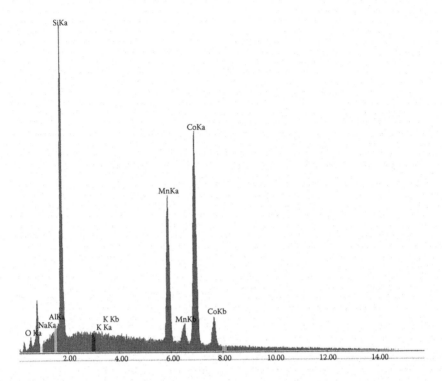

Figure 2.63 Co₂SiMn glass-coated microwires. The glass cover is made of Pyrex and the metallic nucleus has a nominal composition Co₂SiMn. The microwire is made by using Taylor–Ulitovskii technique. (Figure courtesy: Carlos Garcia, from Bogazici University, Department of Physics).

Figure 2.64 Silica nanoparticles of 200 nm (Figure courtesy: Rahmi Ozisik Rensselaer Polytechnic Institute, Department of Materials Science and Engineering).

The ease of sample preparation is one of the reasons of SEM being widely used. Surface charging should also be prevented to obtain a desirable image of the specimen. Electrons from the electron beam accumulate on the surface and create a charged region. In these regions, electron beam that is focused by the probe is deflected in a random way generating artifacts on SEM images. Charging mainly occurs when the specimen is non-conductive. In order to overcome the surface charging, the specimen surface is coated with a conductive material (e.g., gold) by sputtering. In sputtering, the conductive material (gold) is exposed to high-energy ions, which transfer their momentum to the atoms of the conductive material leading them to deposit over the surface of the specimen. For high-magnification imaging, the coating should be thin in order not to lose the details of the surface. Any specimen containing water (i.e., biological specimens) needs further treatment. Water should be removed prior to analysis as the high-energy electron beam evaporates the water into the vacuum, which affects the vacuum system of the microscope. The specimen is dehydrated before the analysis generally by freeze-drying, which is a process where the solvent of a suspension or a solution is frozen and then sublimated under vacuum for removal (Oetjen and Haseley, 2008).

SEMs are applicable to a variety of different specimens and suitable for topographical imaging. Also addition of some other detectors such as EDS improves their range of analysis. With the help of novel advanced computer interfaces, it is relatively easier to operate after suitable training. Despite the need for sample preparation, the time and effort needed are low compared to other similar techniques used for the same purpose. One of the most important advantages of an SEM is that it is capable of working in the magnification range of 20× to greater than 100k× (Leng, 2009). This makes SEM more favorable than optical microscope for low-magnification imaging as illustrated in Figure 2.65, and it can produce surface

(a)   (b)

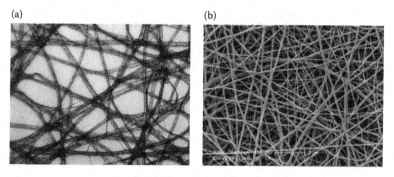

Figure 2.65 Comparison of images of fibers with (a) a light microscope and (b) a scanning electron microscope (Wheeler R., http://www.richardwheeler.net, photograph courtesy: wikipedia).

visualization of an integrated circuit with 3D information in addition to providing the same on top detail obtained in the optical microscope. On the other hand, magnification and resolution limit the application of SEM. If higher magnification and better resolution is needed, TEMs are favored.

Transmission electron microscopy

TEMs are the other type of EMs that use the benefit of shorter wavelength of electrons when compared to light. Although the working principles of SEM and TEM are somewhat similar, type of the specimen and information that is required are important factors in choosing the EM that will be employed for the analysis. Compared to SEM, TEM provides higher magnification and resolution, thus making it suitable for inspecting details of a specimen down to array of atoms.

TEMs are essential analysis instruments in many scientific fields and find application in nanotechnology, materials science, cancer research, and several other fields. TEM image of magnetite nanoparticles is given in Figure 2.66.

The working principle of TEM is quite similar to that of optical microscope. Light generated by a light source is converted into a focused beam by condenser lens and reaches the object. The diffracted light is then focused by an objective lens first to the projection lenses and finally onto the screen as a magnified image of the object. In TEM, the light source is the electron gun providing short-wavelength electrons, the lenses are electromagnetic lenses, and the screen is the fluorescent screen or a CCD (charge-coupled device) camera. The whole path from the electron gun to the screen should be in vacuum as stated before to prevent the undesired scattering of the higher energy electrons. The TEM column is placed

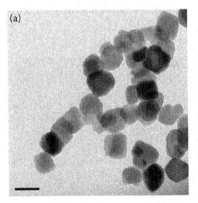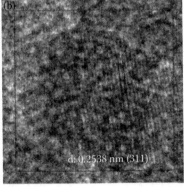

Figure 2.66 Transmission electron microscopy image of magnetite nanoparticles: (a) lattice fringes and (b) d-spacing of 2.54 Å corresponding to (311) plane of magnetite nanoparticles.

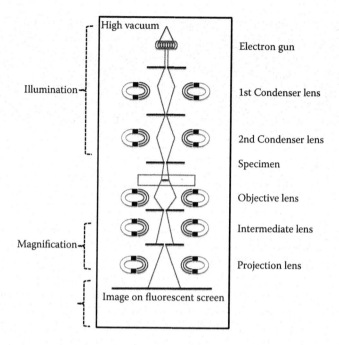

Figure 2.67 Components of transmission electron microscope.

vertically with the electron gun on top and the fluorescent screen and camera at the bottom (Figure 2.67). Modern TEMs are equipped with control pads attached to the microscope for convenient use of the operator and high-end TEMs can even be controlled remotely.

It was previously stated that electromagnetic lenses are used to focus the primary electron beam as glass lenses cannot be used to focus electrons. The main advantage of using electromagnetic lenses over traditional glass lenses is that their magnification limit can be manipulated by changing the electrical current, which affects the magnetic field magnitude (FEI, 2010).

Although in all EMs, electrons are generated by an electron gun, in TEM they are accelerated by high voltage. Better resolution of the TEM is a result of this high voltage, which affects the wavelength of the electrons, thus the final resolution (300 kV voltage can give 0.05-nm resolution).

Sample preparation for TEM is more complicated than SEM or optical microscope. First of all, sample should be very thin in the range 100–500 nm to prevent multiple scattering as often encountered in thick samples. With novel preparation techniques, it is now possible to prepare very thin samples keeping its bulk properties and atomic structures.

The specimen should be prepared as thin as possible for TEM because the interactions between the specimen and primary electrons is highly affected by the sample density as well as the accelerating voltage. As given

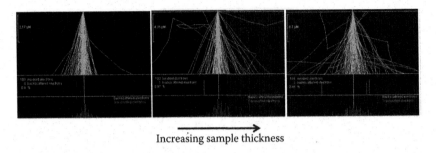

Increasing sample thickness

Figure 2.68 Formation of backscattered electrons on TEM due to increase in sample thickness.

in Figure 2.68, when the thickness of the specimen increases, the amount of backscattered electrons increases, which in turn decreases the amount of electrons that can penetrate into the sample (Goodhew et al., 2012).

A thin layer of the sample is mounted on a mesh disc ($d = 3$ mm) made of copper, which is used to hold the sample. Mesh sizes can be in the range of few to 100 μm. This mesh disc can also be coated with a thin carbon or a polymer film to keep in place the specimen parts that have smaller size than the mesh size of the disc on the mesh. Prepared disc is then mounted on a specific holder, which is inserted into the microscope. This holder is also capable of tilting the mesh disc within the microscope for other purposes such as electron tomography. Biological samples are often embedded into a resin and sliced by microtomy. The sample is then placed onto the mesh grid.

Vitrification, which is the sudden freezing of the water without forming ice crystals, is also used to prepare samples that can be damaged by conventional sample preparation methods or to avoid drying which changes the behavior of particles in solution. Low-dose protocols are generally used to protect the vitrified sample, and the low temperature of the vitrified sample further reduces the damage by the beam during inspection. This allows increasing the exposure time to optimize the image (FEI, 2010). The imaging that is done with vitrified specimen is called Cryo-TEM. The advantage of Cryo-TEM is that it allows the inspection of biological materials or suspended material in solution at their native hydrated state. Cryo-TEM is also used in materials chemistry to image soft materials (e.g., polymers) and crystals at their native environment. The chemical and physical states of some samples may be altered during drying, so time-resolved rapid freezing of the sample can be employed to investigate a dynamic system. TEM is not suitable for all types of samples due to exposure to highly energetic electrons in the presence of high vacuum environment. Especially, samples containing volatile organic compounds or residual solvents may contaminate the column and cause reduction of the image quality (Fahlman, 2011).

As in the case of SEM, there are several possibilities for an electron interaction with the sample within the TEM. A portion of the electrons is absorbed by the specimen proportional to the thickness and atomic structure. This absorption defines the contrast of the non-crystalline material in the image. Specimen containing crystalline material causes scattering of the electrons in a distinct direction, which is called diffraction contrast. A pattern of these diffractions is formed after the objective lens at the back focal plane that is defined as the plane where all the parallel beams leaving the specimen cross over each other and come to a focus. It is possible to project this diffraction pattern to the screen by the help of projection lenses, which are located below the objective lens, and the diffraction pattern can be used to analyze the shape, orientation, and even the type of the specimen by comparing the lattice distances of the diffracted planes with the theoretical data. This technique is known as electron diffraction (ED) and a diffraction pattern of a polycrystalline sample is given in Figure 2.69. Finally, some electrons can be transmitted through the sample and lose their energies according to their interaction with the atomic structure of the sample and are characterized/sorted with respect to this lost energy for elemental and chemical analysis (EELS).

To observe the image, a fluorescent screen is normally used, which is illuminated when in contact with electrons. This screen is also in vacuum and can be inspected by an external window or binoculars. It is also possible to move the screen from the path that the transmitted or diffracted

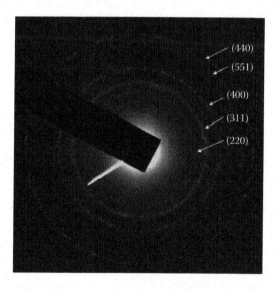

Figure 2.69 Electron diffraction pattern of magnetite nanoparticle. (From Jing, J. et al. *Journal of Nanoparticle Research* 14(4), 2012: 1–8. With permission.)

Figure 2.70 Three-dimensional reconstruction of octahedral magnetite nanoparticle by cryo electron tomography. (Figure courtesy: Cem Levent Altan, Technische Universiteit Eindhoven, Eindhoven, The Netherlands.)

electrons are traveling. This allows the electrons to be projected on a CCD camera.

Although the basic principles are the same, TEMs are different from SEMs in several aspects. In SEM, a dynamic electron beam is used and focused on an area of the object, while a static electron beam is used in TEM. As transmittance is needed for TEM, high voltage is applied to the electrons to make them energetic enough to penetrate into the specimen. Unlike SEM, the samples for TEM should be prepared as thin as possible to allow for electron penetration.

3D imaging is also possible with TEMs by electron tomography. The sample is rotated along an axis with angular increments and then all the corresponding images are acquired. With the help of a computer, the series of the images are collected and merged together to form a 3D tomographic reconstruction of the specimen. Figure 2.70 displays the reconstruction of octahedral magnetite nanoparticles by cryoelectron tomography. The whole process of acquisition can be performed in an automated fashion. This automation can help acquiring thousands of images and also keep the dose at reasonable values in order to reduce sample damage.

References

de Jonge, N. and F.M. Ross. Electron microscopy of specimens in liquid. *Nature Nanotechnology* 6(11), 2011: 695–704.

Dunlap, M. and J.E. Adaskaveg. *Introduction to the Scanning Electron Microscope.* 1997. http://www.geo.umass.edu/courses/geo311/semmanual.pdf

Dürr, M. and U. Höfer. Hydrogen Diffusion on Silicon Surfaces. *Progress in Surface Science* 88(1), 2013: 61–101.

Eaton, P. and P. West. Tip-sample convolution. 2010. http://afmhelp.com/

Fahlman, B.D. *Materials Chemistry*. London: Springer, 2011.

FEI. An Introduction to Electron Microscopy. 2010. http://www.fei.com/resources/student-learning/introduction-to-electron-microscopy/intro.aspx

Frommer, J. Scanning tunneling microscopy and atomic force microscopy in organic chemistry. *Angewandte Chemie International Edition in English* 31(10), 1992: 1298–328.

Goodhew, P., D. Brook, B. Tanovic, A. Green, and I. Jones. Transmission Electron Microscopy (TEM). University of Liverpool. 2012. http://www.matter.org.uk

Hafner, B. Energy Dispersive Spectroscopy on the SEM: A Primer. 2010. http://www.charfac.umn.edu/instruments/eds_on_sem_primer.pdf

Jing, J., Y. Zhang, J. Liang, Q. Zhang, E. Bryant, C. Avendano, V.L. Colvin, et al. One-step reverse precipitation synthesis of water-dispersible superparamagnetic magnetite nanoparticles. *Journal of Nanoparticle Research* 14(4), 2012: 1–8.

Li, A., P. Coombe, and R. Oliver. Techniques for Studying Materials: Atomic Force Microscopy. 2010. University of Cambridge. http://core.materials.ac.uk/repository/doitpoms/tlp/afm.htm

Leng, Y. *Materials Characterization: Introduction to Microscopic and Spectroscopic Methods*. Singapore: Wiley, 2009.

Matsuura, H., H. Furukawa, T. Tanikawa, and M. Ogawa. Dynamics and applications of a microscale liquid surface interacting with an electric Field. *Surface Science* 583(1), 2005: 29–35.

Murphy, D.B. and M.W. Davidson. *Fundamentals of Light Microscopy and Electronic Imaging*. New York, NY: Wiley, 2012.

Oetjen, G.W. and P. Haseley. *Freeze-Drying*. Weinheim, Germany: Wiley, 2008.

van de Leemput, L.E.C., and H. van Kempen. Scanning tunnelling microscopy. *Reports on Progress in Physics* 55(8), 1992: 1165.

Vilalta-Clemente, A. and K. Gloystein. Principles of Atomic Force Microscopy (AFM). 2008. http://www.mansic.eu/documents/PAM1/Frangis.pdf

Wang, J. Scanning Tunneling Microscopy. University of Guelph: University of Guelph, 2006.

Wilson, R.A. and H.A. Bullen. Basic Theory Atomic Force Microscopy (AFM). 2006. http://asdlib.org/onlineArticles/ecourseware/Bullen/SPMModule_BasicTheoryAFM.pdf

chapter three

Colloids and surfaces

Experiment 1: Sedimentation

Purpose

The purpose of this experiment is to determine the size of zeolite particles using the rate of sedimentation and to understand the factors affecting the sedimentation process.

Theoretical background

Sedimentation is a process that arises from Earth's gravitational field. As gravitational force is exerted on particles, unless there is an equal force acting on the particle in the opposite direction, particles do sediment. The rate of sedimentation depends on factors such as the mass of the particle and the medium it travels in. A ball of higher mass sediments faster than a small one, and sedimentation is faster in low-viscosity mediums, such as air, compared to water. However, when particles are small, such as those that are in the colloidal size range, there are forces that act on them apart from the gravitational force. Brownian motion is a continuous random motion that small particles experience. This random motion results in collision of particles and follows a "random walk," in a fashion very similar to that of gas molecules. This movement is responsible for the diffusion of molecules in gaseous or liquid state (Hunter, 1994).

If the size and density of the particles are high enough, then equilibrium may be reached when all particles settle down. On the other hand, for most colloidal systems, it may take a long time for complete sedimentation to take place or the system may reach equilibrium where the number of particles varies with height.

In many cases where colloidal particles are used, centrifugation is necessary to completely sediment the particles. Centrifugal force, F, is equal to

$$F = m\omega^2 r \tag{3.1}$$

where ω is the angular velocity, m the mass of the particle, and r the radius of the path traveled by the practicles.

Centrifugation is a process where the gravitational force acting on the particles is increased by 3–4 orders of magnitude. Particles with a charge or dipole moment can be sedimented by an electric field or by electric field gradient, in processes called electrophoresis and dielectrophoresis, respectively. More information on electrokinetic techniques can be found in Chapter 2.3.

Especially, when the concentration of colloidal particles is high, particles may aggregate during sedimentation. This results in a faster sedimentation of the aggregated particles as they would be acting like larger particles. On the other hand, when particles act like hard spheres with no interaction (usually the case when the particle concentration is low), their sedimentation rate is proportional to the particle size, and sedimentation experiments can be used to determine the particle size or particle size distributions of colloidal suspensions using Stoke's equation.

As shown in Figure 3.1, the overall force balance on the particle is

$$\sum F = F_g - F_b - F_d = \frac{dv}{dt} \tag{3.2}$$

where F_g is the gravitational force acting on the particle and F_d is the drag or frictional force acting on the particle in the direction relative to the fluid flow velocity. The upward force exerted by the fluid is called the buoyancy, F_b. Buoyant force is not a real force that acts on the particles (as in the case of gravitational force), rather a phenomenon that arises due to the difference in pressure exerted on the top and bottom of an object, and is equal to the weight of the displaced fluid. If the particles are falling in the viscous fluid by their own weight due to gravity, as in the case of sedimentation, then a terminal velocity, also known as the settling velocity, is reached when the forces have equilibrated, so that $dv/dt = 0$ and v is constant in Equation 3.2. According to Stoke's equation, the frictional force acting on the interface between the fluid and the particle is related to the viscosity of the fluid, the size of the particle, and the particles' settling velocity:

$$F_d = 6\pi\eta R v_s \tag{3.3}$$

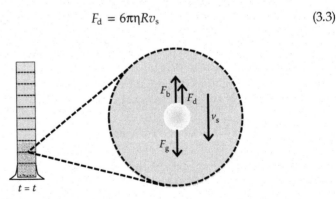

Figure 3.1 The forces acting on an individual particle.

where F_d is the frictional force acting on the interface between the fluid and the particle (N), η the dynamic viscosity (N s m^{-2}), R the radius of the spherical object (m) (particles are assumed to be spherical), and v_s the particles' settling velocity (m s^{-1}).

The settling velocity is given by

$$v_s = \frac{2}{9} \frac{\rho_p - \rho_f}{\eta} gR^2 \tag{3.4}$$

where v_s is the particles' settling velocity (m s^{-1}), g the gravitational acceleration (m s^{-2}), ρ_p is the mass density of the particles (kg m^{-3}), ρ_f is the mass density of the fluid (kg m^{-3}), and η the viscosity of the fluid (N s m^{-2}).

The size of the particles can be determined by measuring the settling velocity. The viscosity of the fluid as well as the density of the medium and the particles can either be measured or literature values can be used. In this experiment, zeolite particles will be sedimented under gravity and the sedimentation rate will be determined by measuring the turbidity of the solutions using a spectrophotometer over a period of time. Using Equation 3.4, the size of the zeolite particles will be determined.

Pre-laboratory questions

1. What is the relation between gravitational, drag, and buoyancy forces acting on a particle?
2. Under what conditions the terminal velocity can be reached?
3. What is the basic difference between centrifugation and sedimentation under gravity?
4. What would you expect for the time required to reach terminal velocity when a more viscous fluid is used?
5. What would you expect for the time required to reach terminal velocity when smaller particles are used?

Chemical and materials

Zeolite particles (6–8 µm) (synthesis given in "Note for the Instructor")
Water
Ethylene glycol ($C_2H_6O_2$)
100 mL of graduated cylinder (to be used as the settling tank)
UV–Vis Spectrophotometer

Procedure

1. Divide the length of the settling tank into 10 equal parts and mark the middle of the top second part (corresponds to 85-mL mark if a

100-mL graduated cylinder is used). A schematic representation of the experiment is given in Figure 3.2.

2. Place 0.1 wt% zeolite particles in 100 mL of water in the settling tank with a mechanical stirrer and stir the solution for 10 min.
3. Stop the mechanical stirrer. Sedimentation under gravitational field starts immediately. Wait for 10 min. This will allow sedimentation of large particles if there are aggregates in solution.
4. After 10 min, take 1 mL of sample using a syringe with a needle where the mark is at 5-min intervals for 30 min ($t = 0$ is 10 min after you stop the stirring).
5. Using a UV–Vis Spectrophotometer, measure the turbidity of each sample at $\lambda = 350$ nm and the original solution immediately after shaking.
6. Repeat the same experiment with zeolite particles suspended in ethylene glycol.

Data and observation

Zeolite particles in water		Zeolite particles in ethylene glycol	
Time (min)	ABS (350 nm)	Time (min)	ABS (350 nm)
0		0	
5		5	
10		10	
15		15	
20		20	
25		25	
30		30	

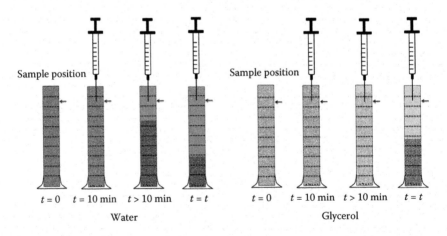

Figure 3.2 Sampling during sedimentation.

1. Plot absorbance at $\lambda = 350$ nm versus time.
2. Obtain a straight line using least-squares fit and extrapolate time to zero absorbance. This is Δt.
3. The distance from the top of the liquid level in the settling tank to the mark where sample is being taken is Δx. $\Delta x / \Delta t$ is the settling velocity v_s.
4. From the obtained settling velocity, using Stoke's equation, determine the size of your particles. For the purpose of calculation, use the density and viscosity values for the solvent from the literature.

Clean-up and disposal

1. Collect the solutions in separate collection bottles.
2. Clean all the glassware and laboratory station.
3. Wash your hands thoroughly before leaving the laboratory.

Post-laboratory questions

1. Why do you think the size you obtained from the sedimentation experiment does not exactly correspond to the stated size of your zeolite particles?
2. Explain the reasons for different settling velocities for particles in water and ethylene glycol.
3. What do you think can be the sources of error in this experiment and what can you do to minimize them?
4. If you were to use a higher concentration of zeolite particles, how would you expect this to affect your accuracy? Why?

Note for the instructor

This experiment can be performed with particles other than zeolite as well (e.g., silica). However, the size of the particles should be in the 1–10 μm range. When very small particles are used, sedimentation is very slow to take measurements over a period of a laboratory hour (i.e., 2–3 h). However, if small particles are to be used, centrifugation of the sample will be required and the g value should be inserted into the equation accordingly. Different particles with different sizes can also be used to immediately see the effect of size on the sedimentation velocity. If students take samples from the middle of their settling tank, they will see no change in the concentration for a period of time and then a decrease. Some students may do this sampling to show the difference. Zeolite particles can be prepared in advance for the following the procedure (Hagiwara et al., 1988):

Synthesis of zeolite particles

1. Place 13.72 g of sodium aluminate ($Na_2O \cdot Al_2O_3$) and 45 g of sodium hydroxide (NaOH) in 400 mL of distilled water in an Erlenmeyer flask and shake to dissolve.

2. Place 123.32 g of sodium metasilicate ($Na_2SiO_3 \cdot 5H_2O$) and 417.4 g of distilled water in another Erlenmeyer flask and shake to dissolve.
3. Mix the two solutions until you obtain a homogenous mixture. An increase in turbidity and viscosity will be apparent. Place the solution in an oven at 90°C for 3 days.
4. After 3 days, separate the zeolite particles from the medium using vacuum filtration.
5. Dry the filtrate for a day in an oven at 90°C.
6. Pulverize the filtrate using a mortar and pestle.

Reference

Hunter, R.J. *Introduction to Modern Colloid Science*. Oxford: Oxford University Press, 1994.

Hagiwara, Z., Hoshino, S., Ishino, H., Nohara, S., Tagawa, K., and Yamanaka, K. Polymer particle having an antibacterial property containing zeolite particles therein and the processes for producing same. US Patent US4775585A Polymer, 1988.

Experiment 2: Determination of critical micelle concentration of surfactants and factors affecting the critical micelle concentration

Purpose

In this experiment, the surface tension and electrical conductivity of various surfactants (SDS, DTAB, CTAB, and $C_{12}E_5$) will be investigated to determine their critical micelle concentration (CMC). Factors such as nature of the surfactant, temperature, and addition of a cosolvent affecting the CMC will be explored.

Theoretical background

Surfactant is an abbreviation for surface-active agent, which includes molecules that are active at surfaces. These molecules have a tendency to reside rather at the surface than in bulk solutions due to their amphiphilic nature. Amphiphilic molecules consist of at least two parts: one of which being hydrophobic and the other hydrophilic. A typical surfactant molecule consists of a long hydrocarbon "tail" that dissolves in hydrocarbon and other nonpolar solvents (water-insoluble; hydrophobic), and a "headgroup" that dissolves in polar solvents (typically water; hydrophilic) as shown in Figure 3.3 (Holmberg et al., 2002). Surfactants have two main features making them essential; one of them is the tendency to adsorb at interfaces and lower the surface tension and the other is the association in solution.

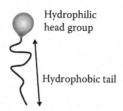

Hydrophilic
head group

Hydrophobic tail

Figure 3.3 Schematic illustration of a surfactant.

One of the fundamental properties of surface-active agents is the self-assembly of surfactant molecules in the bulk solutions to form aggregates with different geometries (disks, spheres, cylinders, etc.). The first-formed or the simplest aggregates are generally spherical in shape and are called micelles. However, this phenomenon occurs only when the surfactant concentration exceeds a threshold known as CMC. In a micelle, the hydrophobic part of the surfactant molecule is directed toward the interior of the cluster and the polar headgroup toward the aqueous solution (Figure 3.4a). When a surfactant adsorbs from aqueous solution at a hydrophobic surface, it orients its hydrophobic group toward the surface and exposes its polar group to water (Figure 3.4b). The driving force for self-assembly is said to be the hydrophobic effect (Tanford, 1973). As micelles form (Figure 3.4c), sharp changes occur in many physical properties such as the surface tension, viscosity, conductivity, and sometimes turbidity of the solution.

When a surfactant is added to an aqueous solution, it first absorbs at the surface with its headgroup in the liquid and the tails sticking out away from the aqueous phase. At these low concentrations, below the CMC, only unimers exist in the solution and increasing the surfactant concentration in the bulk aqueous phase also results in increasing occupancy of the surface (i.e., increasing surface concentration). Ultimately, when the surface is

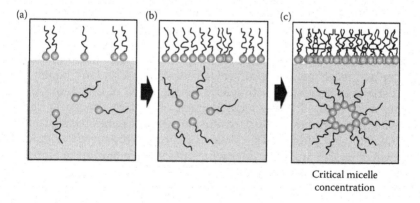

(a)   (b)   (c)

Critical micelle
concentration

Figure 3.4 Surfactant behavior in aqueous solutions.

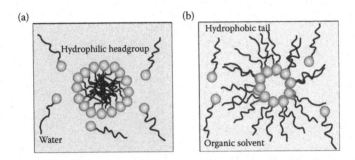

Figure 3.5 Spherical micelle representations: (a) in aqueous solution and (b) in organic solvent.

fully covered, unimers start to increase in number inside the solution. As the CMC is reached, micelles start forming, being in equilibrium with the unimers as shown in Figure 3.5. Above the CMC, the concentration of the surfactant molecules at the interface remains constant by dynamic equilibrium and only the number of micelles in solution increases.

As discussed above, the other characteristic feature of surfactants is their tendency to accumulate at surfaces to lower the free energy of the system. Surface tension is the free energy per unit area of the boundary between water and the air above it. Surface tension will be reduced as this boundary gets increasingly populated by surfactant molecules. The denser the surfactant packing at the surface, the larger will be the reduction in surface tension.

Surface tension (γ) is the force along a line of unit length, where the force is parallel to the surface but perpendicular to the line. Its SI unit is Newton per meter (N m^{-1}). Another definition of surface tension is the amount of work done to create a unit surface area. In order to increase the surface area of a mass of liquid by an amount, work, w, is needed. This work is stored as potential energy. Therefore, surface tension can also be expressed in joules per square meter (J m^{-2}). In its quest to find a state of minimum potential energy, a droplet of liquid spontaneously assumes a spherical shape, which has the minimum surface area for a given volume. More details on surface tension and measurement of surface tension can be found in Chapter 2.1.

Various methods exist to determine the CMC of a system. A commonly used method is the measurement of surface tension at various surfactant concentrations. Surface tension decreases until the surface gets saturated (i.e., no additional surfactant can go onto the surface without the removal of an existing one), after which the surface tension remains constant. Further addition of surfactants to the solution leads to micellization, and the CMC point can be located as represented in Figure 3.6.

For ionic surfactants, another method that can be used to determine the CMC is the measurement of electrical conductivity. Electrical conductivity

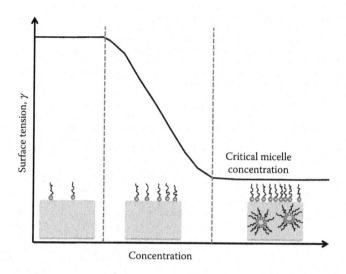

Figure 3.6 Surface tension vs. concentration graph.

is a measure of a material's ability to conduct an electric current. When there is an electrical potential difference across a conductor, an electric current occurs as a result of the flow of movable charges. The conductivity of an aqueous solution is highly dependent on its concentration of dissolved salts and sometimes other chemical species, which tend to ionize in solution. Below the CMC, the addition of a surfactant to an aqueous solution causes an increase in the number of charge carriers (Na^+(aq) and $C_{12}H_{25}OSO_3^-$(aq) for SDS) and consequently, an increase in the conductivity. Above the CMC, further addition of surfactants increases the micelle concentration, while the monomer concentration remains approximately constant (at the CMC level). Since a micelle is much larger than a unimer, it diffuses more slowly through solution and so is a less efficient charge carrier. When conductivity is plotted against surfactant concentration, a change in the slope of the linear relationship is observed at the CMC point as shown in Figure 3.7.

There are several factors that affect the CMC. Briefly, these are:

1. The length of the alkyl group: Adding a $-CH_2$ group lowers the CMC by about a factor of 5, regardless of the headgroup.
2. Nature of the headgroup: Nonionic surfactants have much lower CMC values than ionic surfactants.
3. Temperature: For ionic surfactants, temperature does not have much of an effect. In nonionic systems, CMC decreases with increasing temperature.
4. Addition of an electrolyte (only for ionic surfactants): CMC decreases as electrolyte concentration increases.

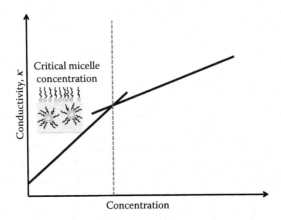

Figure 3.7 The change in electrical conductivity with respect to concentration.

Further information on factors affecting the SMC can be found elsewhere (Holmberg et al., 2002). In this experiment, CMC of different surfactants will be determined for an anionic (SDS), two cationics (DTAB, CTAB), and a nonionic ($C_{12}E_5$) surfactant. In order to investigate the factors affecting the CMC, the effect of temperature, ionic strength, and the addition of a solvent will be explored for the aforementioned surfactants.

Pre-laboratory questions

1. What are the factors affecting the critical micelle concentration (CMC)?
2. Why does surface tension remain constant once CMC is reached?
3. Can conductivity be used to measure CMC for a nonionic surfactant and why?
4. Read the safety data for the surfactants SDS, CTAB, and $C_{12}E_5$.

Materials

Sodium dodecyl sulfate (SDS)
Dodecyl trimethyl ammonium bromide (DTAB)
Cetyl trimethyl ammonium bromide (CTAB)
Dodecyl pentaethylene glycol ether ($C_{12}E_5$)
Ethanol (C_2H_5OH)
Sodium chloride (NaCl)

Procedure

Determination of CMC of surfactants by surface tension/electrical conductivity
1. Prepare 100 mL of 50 mM SDS, 100 mM DTAB, 10 mM CTAB, and 0.1 mM $C_{12}E_5$ stock solutions in water.

2. Using the stock solutions, prepare a series of solutions by the following dilution factors: 2×, 4×, 5×, 6×, 7×, 10×, 20×, 50×, 100×.
3. Measure the surface tension/electrical conductivity of all prepared solutions with the help of the instructions manual for surface tension meter/conductivity meter and record in data table.
4. Plot the surface tension/electrical conductivity versus concentration and determine the CMC value.

Determination of the effect of temperature on CMC

1. Measure the surface tension of $C_{12}E_5$ solutions at 50°C using a surface tension meter and record the results in the data table.
2. Plot the surface tension versus concentration and determine the CMC value.

Determination of the effect of electrolyte addition on CMC

1. Prepare 1 L of 0.05 M NaCl stock solution.
2. Using the surfactant stock solutions and the NaCl stock solution, prepare the same dilutions of DTAB solutions with a final NaCl concentration of 1 mM.
3. Measure the surface tension/electrical conductivity of all prepared solutions with the help of the instructions manual for surface tension meter/conductivity meter and record the data in the table.
4. Plot the surface tension/electrical conductivity versus concentration and determine the CMC value.

Determination of the effect of solvent addition on CMC

1. Prepare 1 L of 20:80 ethanol:water stock solution.
2. Prepare new surfactant stock solutions in 100 mL of ethanol:water solution. Using this, prepare the same dilutions of DTAB solutions using 20:80 ethanol:water solution as the solvent.
3. Measure the surface tension/electrical conductivity of all prepared solutions with the help of the instructions manual for surface tension meter/conductivity meter and record the data in the table.
4. Plot the surface tension/electrical conductivity versus concentration and determine the CMC value.

Clean-up and disposal

1. Collect the solutions in separate collection bottles.
2. Clean all the glassware and laboratory station.
3. Wash your hands thoroughly before leaving the laboratory.

Observations and data

Surface tension

SDS		DTAB	
Concentration (mM)	Surface tension (mN m⁻¹)	Concentration (mM)	Surface tension (mN m⁻¹)
—			

CTAB		$C_{12}E_5$	
Concentration (mM)	Surface tension (mN m⁻¹)	Concentration (mM)	Surface tension (mN m⁻¹)

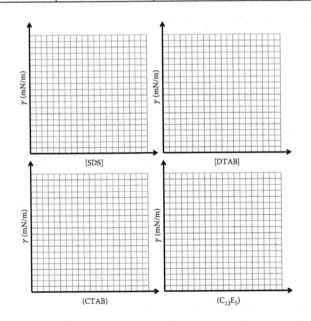

Electrical conductivity

SDS		DTAB	
Concentration (mM)	Electrical conductivity (S m^{-1})	Concentration (mM)	Electrical conductivity (S m^{-1})
CTAB			
Concentration (mM)	Electrical conductivity (S m^{-1})		

DTAB at 50°C		DTAB in 20:80 ethanol:water	
Concentration (mM)	Electrical conductivity (S m^{-1})	Concentration (mM)	Electrical conductivity (S m^{-1})

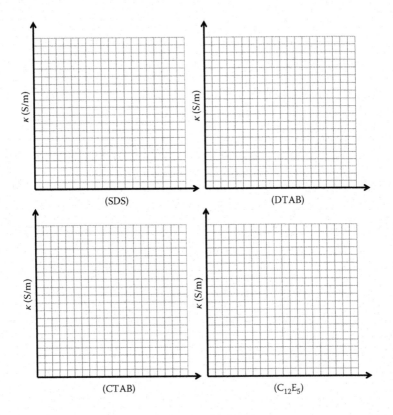

Post-laboratory questions

1. Compare your CMC results with those in the literature and explain discrepancies, if any. Discuss possible sources of error. How would you do the experiment differently?
2. How does the length of the hydrophobic chain affect the CMC? Can you make a generalization?
3. How does the nature of the headgroup affect the CMC? Can you make a generalization?
4. Why do electrolytes affect the CMC so strongly?
5. Explain the effect of temperature on CMC.
6. How does addition of a solvent affect the CMC of a surfactant?
7. Suggest other experiments to determine the CMC of a surfactant by considering the factors affecting the CMC.

Note for the instructor

For a large group of students, surface tension measurements can be done most practically using a conductivity meter. If this is the case, only ionic surfactants can be used. Any surfactant can be used in this experiment; however, the concentration of the solutions should be adjusted to measure a few concentrations below and above the CMC. If surface tension is to be measured with a drop-shape analyzer, then it is sufficient to make 5 mL of surfactant solutions. For conductivity, 10–20 mL of each solution is sufficient. If duNuoy ring or Wilhelmy plate is to be used, approximately 50 mL of each solution is required. The temperature effect cannot be investigated with a drop-shape analyzer unless there is a special temperature cabinet for the instrument. Because there are several sections to this experiment, different groups can investigate different sections and share data to optimize the time.

References

Holmberg, K., B. Jonsson, B. Kronberg, and B. Lindman. *Surfactants and Polymers in Aqueous Solution*. 2nd Edition. Chichester: Wiley, 2002.
Tanford, C. *The Hydrophobic Effect: Formation of Micelles and Biological Membranes*. New York: John Wiley & Sons, 1973.

Experiment 3: Adsorption at interfaces: Gibbs adsorption isotherm

Purpose

The purpose of this experiment is to measure the surface tension of a non-ionic surfactant and use the Gibbs adsorption equation to determine the adsorption isotherm. Then, dynamic surface tension of the same surfactant at different concentrations and temperatures is measured and data are interpreted in terms of an appropriate adsorption model.

Theoretical background

Equilibrium surface tension

General information on equilibrium surface tension can be found in Experiment 2: Determination of CMC of surfactants and factors affecting the CMC and in surface tension measurements given in Chapter 2.1.

Adsorption

As surface tension, γ, is lowered when molecules adsorb at the liquid–air interface, the extent to which surface tension is lowered is a measure of the adsorbed amount and this relation is given by Gibbs adsorption equation. Surface concentration (Γ) is defined as the number of solute molecules at the surface per unit area. Once equilibrium surface tension is measured, the surface concentration can be calculated using the Gibbs equation

$$\Gamma = -\frac{1}{nRT}\frac{d\gamma}{d\,ln(c)} \tag{3.5}$$

where Γ is the equilibrium surface excess (surface concentration), R the gas constant, T the temperature in Kelvin, and c is the bulk surfactant concentration. $n = 1$ for nonionic surfactants, neutral molecules, or ionic surfactants in the presence of excess electrolyte. Adsorption isotherm is achieved by plotting surface excess (Γ) versus bulk surfactant concentration (c).

Dynamic surface tension

Although equilibrium surface tension measurements are useful in the determination of adsorption isotherms, equilibrium surface tension is not instantaneously achieved. When a surfactant molecule is first introduced into the solution, these molecules must diffuse to the surface and then adsorb with the correct orientation. At equilibrium, molecules adsorb and desorb at the same rate. Once at equilibrium, if the surface is stretched, surfactant molecules will transport from the bulk into the surface until equilibrium is reached. If the surface is contracted, then some of the surfactant molecules will transport back into the bulk, reestablishing the equilibrium. There are two main models that attempt to describe this transport and adsorption/desorption processes (Ravera et al., 1993; Liggieri et al., 1996).

The mechanism of adsorption can be diffusion-controlled where the surfactant molecules are assumed to diffuse to an imaginary subsurface and once there, adsorb at the surface. In this mechanism, the rate-controlling step is the diffusion of monomers from the bulk to the subsurface.

In the other model, the adsorption of molecules from the subsurface to the surface is thought to be the rate-determining step. Once the monomer is at the subsurface, the adsorption may be hindered by an increased surface pressure, the presence of less vacant sites, or by steric constraints that prevent the monomer to adsorb with the correct orientation. These may result in the monomer to back diffuse into the bulk and more time may be required for the equilibrium surface tension to be achieved. This model is the mixed diffusion-kinetic-controlled adsorption. Once the monomers diffuse to the imaginary subsurface, only those that possess energy higher than an activation barrier can be adsorbed at the surface. Especially in the case of larger macromolecules such as polymers and proteins, the orientation of adsorption also becomes a factor in adsorption. At the later stages of adsorption when most sites are occupied, the monomers in the bulk statistically will have less of a chance to strike an empty space. None of these aforementioned mechanisms can take place alone and each will contribute to a delay in the adsorption compared to the diffusion-only model. Measuring dynamic surface tension with respect to concentration and temperature gives information about the adsorption mechanism. Delayed adsorption suggests the mixed model to predominate, and temperature dependence of dynamic surface tension supports the activation energy suggestion.

The time dependence of adsorption in the long time limit is given as

$$\gamma(t)_{t\to\infty} = \gamma_{eq} + \frac{RT\Gamma^2}{2c}\left(\frac{\pi}{Dt}\right)^{1/2} \tag{3.6}$$

where γ_{eq} is the equilibrium surface tension, c the bulk concentration, and D the monomer diffusion coefficient of the surfactant. D is experimentally determined, usually using NMR and for the nonionic surfactant Di(C6-Glu), that will be used in the experiment D has been determined to be 2.7×10^{-10} m^2 s^{-1} at 25°C (Griffiths et al., 1997).

A mixed diffusion-kinetic model, which describes the surface adsorption kinetics, assumes that transport of surfactant monomer from the subsurface to the interface is the rate-determining step and an adsorption barrier hampers adsorption. This activation energy barrier E_a is related to the effective diffusion coefficient (D_{eff}) (Eastoe et al., 1998). Measuring dynamic surface tension allows for the determination of an effective diffusion coefficient, D_{eff}, and the D_{eff}/D ratio gives information about the mechanism of adsorption. If the mass transport to the interface is explained only by Fickian diffusion, then $D_{eff}/D = 1$. However, for dilute solutions, where the concentration is less than the CMC, mixed diffusion-kinetic adsorption mechanism is established. When the monomer diffuses to the subsurface, its adsorption is prevented due to the activation energy barrier, E_a. In this case, it is more favorable to go back to the bulk rather than adsorption, leading to an increase in the time for surface tension decay ($D_{eff}/D < 1$).

In this experiment, equilibrium surface tension of a nonionic surfactant will be measured and Equation 3.5 will be used to calculate the surface concentration to obtain Gibbs adsorption isotherm. Dynamic surface tension measurements will be done at different temperatures to understand the surfactant adsorption mechanism at the water–air interface.

Pre-laboratory questions

1. What is the difference between equilibrium and dynamic surface tension?

2. Under what condition the diffusion of monomers from the bulk is the rate-controlling step of adsorption?
3. What is the effect of surface pressure on adsorption?

Materials

Di(C6-Glu) $((C_6H_{13})_2C[CH_2NHCO(CHOH)_4CH_2OH]_2)$
Water

Procedure

Gibbs adsorption isotherm
1. Prepare 2, 5, 10, 12, 15, 20, and 25 ($\times 10^{-4}$) M Di(C6-Glu) solutions and measure the equilibrium surface tension at 25°C.
2. Plot the surface tension versus concentration data and determine the CMC from the point where the surface tension remains constant.
3. Using the Gibbs adsorption equation (Equation 3.5), calculate surface excess (Γ) and plot surface excess (Γ) versus bulk surfactant concentration (c) to obtain the Gibbs adsorption isotherm.

Dynamic surface tension
1. Take the 5×10^{-4} mol dm^{-3} Di(C6-Glu) solution, measure the dynamic surface tension at 10, 20, 30, 40, and 50°C at 0.001 s (1 ms) time intervals for 10 s.

Data and observations

1. Plot dynamic surface tension versus time (time in log scale) for all temperatures on the same graph.
2. Plot dynamic surface tension versus (time)$^{-1/2}$ for all temperatures on the same graph.
3. Determine the linear regime for longer times and obtain straight lines using least-squares fit.
4. Determine the slope of each straight line.
5. Determine the effective diffusion coefficient, D_{eff}, using

$$D_{eff} = \left(\frac{RT\Gamma^2\pi^{(1/2)}}{2c(\text{slope})} \right)^2$$

(3.7)

6. Calculate D of the monomer at different temperatures using $D = 2.7 \times 10^{-10}$ m^2 s^{-1} at 25°C and Stokes–Einstein equation:

$$D = \frac{k_B T}{6\pi\eta R_H} \tag{3.8}$$

where R_H is the radius of the spherical particle, k_B the Boltzmann constant, η the dynamic viscosity, and T the temperature.
7. Calculate D_{eff}/D.

Concentration, c (×10⁻⁴ M)	γ(mN m⁻¹)	Γ (mol m⁻²)	
2			
5			
10			
12			
15			
20			
25			

T/°C	Slope/mN · m⁻¹s¹/²	10¹⁰ D_{eff}/m² · s⁻¹	10¹⁰ D/m² · s⁻¹	D_{eff}/D

Clean-up and disposal

1. Discard the solutions.
2. Clean all the glassware and laboratory station.
3. Wash your hands thoroughly before leaving the laboratory.

Post-laboratory questions

1. How would you explain the shape of the adsorption isotherm?
2. How does dynamic surface tension change with temperature? What does this signify in terms of activation energy of monomer adsorption from the subsurface?
3. What are the two mechanisms acting on adsorption as suggested by D_{eff}/D ratio? What does it signify when D_{eff}/D approaches 1?

Note for the instructor

If dynamic surface tension measurements are to be performed, this can be done with any nonionic surfactant. If only equilibrium surface tension is to be measured, Gibbs adsorption isotherms can be drawn for any surfactant system. Therefore, this experiment can be combined with Experiment 2, and surface tension data obtained in Experiment 2 can be used to draw the Gibbs adsorption isotherms. This is a lengthy experiment. Measurement of dynamic surface tension at different temperatures can be performed by different groups and data can be shared.

References

Eastoe, J., J.S. Dalton, P.G.A. Rogueda, and P.C. Griffiths. Evidence for activation–diffusion controlled dynamic surface tension with a nonionic surfactant. *Langmuir* 14(5), 1998: 979–81.

Griffiths, P.C., P. Stilbs, K. Paulsen, A.M. Howe, and A.R. Pitt. FT-PGSE NMR study of mixed micellization of an anionic and a sugar-based nonionic surfactant. *Journal of Physical Chemistry B* 101(6), 1997: 915–18.

Liggieri, L., F. Ravera, and A. Passerone. A diffusion-based approach to mixed adsorption kinetics. *Colloids and Surfaces A: Physicochemical and Engineering Aspects* 114, 1996: 351–59.

Ravera, F., L. Liggieri, and A. Steinchen. Sorption kinetics considered as a renormalized diffusion process. *Journal of Colloid and Interface Science* 156(1), 1993: 109–16.

Experiment 4: Exploration of wetting phenomena

Purpose

The purpose of this experiment is to develop an understanding of the relationship between the contact angle and the wettability of solid surfaces. Roughness of the solid surface changes the contact angle of a liquid with that solid, affecting its wettability. Zisman plot will be used to determine the critical surface tension of solid surfaces with different roughness.

Theoretical background

When a drop of insoluble liquid comes into contact with a solid surface or another liquid, the drop assumes a particular shape depending on the surface tension of the air–liquid–solid interface. Some drops are more "round," suggesting a dislike for the surface, whereas "flat" drops suggest a liking towards the surface (Hiemenz and Rajagopalan, 1997).

A drop may act in three different ways when placed on a surface (Figure 3.8):

1. The liquid wets the surface and spreads across the surface (complete wetting).
2. The liquid spreads over the surface as a thin film and the excess remains as a drop on the surface (moderate wetting).

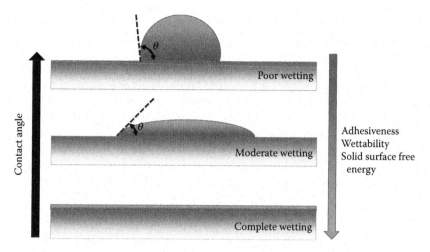

Figure 3.8 The representation of a drop on different surfaces.

3. The liquid does not wet the surface and the drop remains intact. Over a period of time, some spreading may occur due to solubility (poor wetting).

Whatever the tendency of the drop be in terms of wetting the surface, it is subject to change as spreading changes the initial surface tension. Therefore, the initial and later stages of spreading may differ.

Surface tension and the contact angle are related to each other by Young's equation. From Figure 3.9, force balance in the horizontal direction results in the following equality:

$$\gamma_{LV} \cos\theta = \gamma_{SV} - \gamma_{SL} \tag{3.9}$$

where γ_{LV} is the surface tension of the liquid–vapor interface and γ_{SV} and γ_{SL} represent the solid–vapor and solid–liquid interfaces, respectively.

The interfacial tension can be determined by measuring the contact angle, θ, and the surface tensions can be obtained. The equilibrium need not be between vapor, liquid, and solid only, but it can also be between two liquids and a solid, in which case the equation should be altered

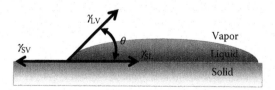

Figure 3.9 Young's equation is used to describe the interactions between the forces of cohesion and adhesion and measure what is referred to as surface energy.

accordingly. This equation assumes the surface to be perfectly homogenous, without any surface roughness or deformability.

Measurement of the contact angle gives us information about the wettability of the surfaces. For some applications in industry, complete wetting is required, whereas for some applications minimizing the contact angle is the purpose. As the drop wets the surface, the contact angle approaches zero. Controlling the contact angle and being able to tune it for the required application is possible upon addition of surface-active agents. Surface roughness plays an important role in the contact angle. Rough surfaces have high contact angles as spreading becomes easier.

The critical surface tension of a material, γ_c, is a measure of the surface's wettability and it is proportional to the surface free energy of the material. A liquid with a surface tension less than or equal to the critical surface tension of a particular material will "wet" that surface, that is, the contact angle will be less than or equal to 90°. Zisman noticed empirically that a plot of $\cos\theta$ versus γ_{LV} is often linear (Fox and Zisman, 1950). Therefore, a material's critical surface tension can be determined from a Zisman plot, which measures variation in Young's contact angle as a function of the surface tension of a series of liquids. The value for which $\cos\theta$ extrapolates to 1 is termed the critical surface tension.

In this experiment, the contact angle between a substrate and the number of solvents will be measured, and Zisman plot will be generated to determine the critical surface tension of the substrate. The same substrate will be polished with sandpaper to change its surface roughness, resulting in a change in the critical surface tension (Nilsson et al., 2010).

Pre-laboratory questions

1. What do low values of θ suggest about a liquid's interaction with a surface?
2. What is critical surface tension? How do you expect this to change with surface roughness?
3. How would you expect the contact angle of water on a Teflon surface to change upon addition of a water-soluble surfactant?
4. What do you expect to see if the liquid surface tension is less than or equal to the critical surface tension of a surface?
5. Give an example of a hydrophobic and hydrophilic surface encountered in everyday life.

Chemicals and materials

Water
Ethylene glycol ($C_2H_6O_2$)
Formamide (CH_3NO)
Dichloromethane (CH_2Cl_2)

Acetone (C_3H_6O), for cleaning
Ethanol C_2H_5OH, for cleaning
1-mL syringe
2 flat pieces of polytetrafluoroethylene (PTFE; commercial name: Teflon)
   (~3 cm × 3 cm) 150 grit sand paper

Procedure

Preparation of solid surfaces

1. Take two pieces of PTFE surfaces and using the 150 grit sand paper, sand one of the PTFE surfaces randomly in all directions for about 20 s in each direction.
2. Clean both the PTFE surfaces with acetone and then with distilled water. Dry it with pressurized air.
3. Place PTFE (one piece at a time) on the sample stage.

Preparation of liquid

1. Clean syringe by filling it with acetone and emptying three times.
2. Pre-rinse the syringe with the testing liquid.
3. Fill the syringe with ~0.1 mL of testing liquid and secure in the syringe holder above the sample stage.

Instrumentation

1. Turn on the light switch of the sample stage.
2. Adjust the height of the syringe so that the tip can be seen at the top of the screen. Center the tip of the syringe on the screen.
3. Place the PTFE sample on the stage and make sure that the solid sample is flat.
4. Adjust the height of the solid sample. The solid surface should be visible at the bottom of the screen and there should be a gap between the tip of the syringe needle and the solid surface.

Measurement

1. Slowly turn the syringe plunger until a drop falls from the syringe onto the surface. In most instruments, a drop dispenser automatically delivers the drop to the surface. Adjust the focus. Make sure that you can see where the drop meets its reflection on the solid surface. Allow the drop to equilibrate on the sample stage for about 10–30 s.
2. Record the image. An image of the picture taken by the camera should appear on the computer screen. Manually adjust the baseline until the tangent lines look accurate. The instrument calculates the contact angle.
3. If the contact angle is not automatically calculated, create a new drop on an un-wetted part of the surface.

4. Repeat the process for a total of five measurements for this liquid. Before each measurement, the solid surface should be cleaned and dried. Record your contact angle values.
5. After five measurements, empty the syringe into the waste and begin the procedure again for the remaining liquids and record your contact angle values. Between different test liquids, the solid surface and the syringe should be cleaned as described previously.
6. Measure the contact angles for all liquids for the sanded PTFE sample as described above.

Observations and data

		θ	$\cos\theta$	γ_{LV} (mN m^{-1})
PTFE	Water			72.8
	Formamide			58.20
	Ethylene glycol			47.7
	Dichloromethane			26.5
PTFE with 150-grid sand paper	Water			72.8
	Formamide			58.2
	Ethylene glycol			47.7
	Dichloromethane			26.5

1. Find the mean $\cos\theta_\gamma$ for each liquid and the standard deviation of $\cos\theta_\gamma$.
2. Plot the mean cosine value for each liquid against the surface tension of the liquids (Zisman plot). Find the best linear fit and extrapolate this line to the cosine value of 1 to find the surface tension value of the solid–gas interface (γ_c).

Clean-up and disposal

1. Collect the solutions in separate collection bottles.
2. Clean all the glassware and laboratory station.
3. Wash your hands thoroughly before leaving the laboratory.

Post-laboratory questions

1. Explain the differences you observed in the contact angles for the different liquids. Do you see a trend?
2. Compare your calculated value of the surface tension of untreated PTFE to a literature value. Explain any discrepancies.
3. How does surface roughness affect the wettability?
4. How can the Zisman plot be used to estimate the surface free energy of a solid? What assumptions are made in estimating this value?

Note for the instructor

In this experiment, instead of investigating the effect of surface roughness, different solid substrates can also be used to show how wetting changes when the nature of the substrate changes. This will allow for the measurement of γ_c for different solid substrates. If the contact angle is too low (the liquid wets the surface), it is not possible to make facile measurements, so substrate/liquid couple should be chosen accordingly. For the solvents used in the experiment, a hydrophobic surface is more appropriate. Due to time constraints, each group may determine γ_c for a particular surface only. If there is no contact angle measurement device in the laboratory, photographs of the drop on the surface can be taken under magnifying glass and the θ values can be determined from the images.

References

Fox, H.W. and W.A. Zisman. The spreading of liquids on low energy surfaces. I. Polytetrafluoroethylene. *Journal of Colloid Science* 5(6), 1950: 514–31.

Hiemenz, P.C. and R. Rajagopalan. *Principles of Colloid and Surface Chemistry*. 3rd Edition. New York: Marcel Dekker, 1997.

Nilsson, M.A., J.R. Daniello, and P.J. Rothstein. A novel and inexpensive technique for creating superhydrophobic surfaces using Teflon and sandpaper. *Journal of Physics D: Applied Physics* 43(4), 2010: 045301.

Experiment 5: Determination of polymer shape in solution using viscosity measurements

Purpose

The purpose of this experiment is to measure the viscosities of the same polymer of different molecular weights in different solvents, which can be good, poor, or theta solvents at certain temperatures. Mark–Houwink plot will be used to determine the shape of the polymer in different solvents.

Theoretical background

Polymers are composed of a number of repeating monomer units. Depending on the number of repeating units, polymers have different molecular weights. These molecules may have either one of these three architectures: linear, branched, or cross-linked. Polymers in solution behave very differently than in the solid state as they are surrounded with solvent molecules. A cross-linked polymer cannot be dissolved; instead, it can only be swollen in a solvent.

Every polymer spontaneously assumes a particular shape, called configuration, when placed in a solvent. A fully stretched conformation is highly unlikely; however, the distance between the ends of a fully stretched conformation is called the contour length, which is proportional to the molecular weight of the polymer. Often, polymer assumes a random coil conformation.

Because polymer is a macromolecule and monomers are connected to each other, polymers behave differently to small discrete molecules. Two monomers can neither occupy the same space nor overlap. This results in excluded volume where one part of a long-chain molecule cannot occupy space that is already occupied by another part of the same molecule.

Polymer molecules move in random walk, and the excluded volume concept results in a self-avoiding random walk. This is a movement that resembles Brownian motion of small molecules in solution. The monomers of the polymer chain assume conformations based on random walk, which results in a random coil. Roughly, this results in a spherical shape where the radius of that sphere is called the radius of gyration, R_g (Figure 3.10).

There are different degrees to the flexibility of a polymer in solution. This is defined with persistence length. The persistence length can be thought of as the longest distance a part of the polymer remains rigid. These can be visualized as the "steps" of the random walk.

At low concentrations, polymers are separated from each other. As the concentration increases, polymer chains begin to touch each other, and at a concentration defined as the overlap concentration, c^*, the whole volume of solution is visualized as packed spheres. Below the overlap concentration, polymer solutions are considered dilute where polymers behave independently from each other and predominantly interact with solvent molecules. At concentrations above c^*, the solution is in the semi-dilute regime where the chains are overlapped and entangled. Their mobility is greatly reduced. At even higher concentrations, concentrated regime starts where each segment of the polymer chain does not have enough space available. The overlap concentration depends on R_g. The concentration regimes are different for rigid-chain molecules. Compared with a linear flexible chain, c^* is much lower.

Polymer behavior in solution depends very much on the solvent. If the solvent is good, meaning that the interactions between the solvent and

Figure 3.10 Self-avoiding random walk.

the polymer are favorable, then the polymer extends out into the solution. However, in a poor solvent where the interactions are not favorable, the polymer–polymer interactions will be preferred resulting in the shrinking of the polymer. Whether a solvent is good or poor depends on the solvent–polymer pair and on the temperature (Hunter, 1994).

In poor solvents, the effective attraction between monomers causes neutral polymer chains to collapse into dense spherical globules in order to maximize the number of favorable polymer–polymer interactions and minimize the unfavorable polymer–solvent contacts. This is somewhat similar to micellization. Here, instead of separate surfactant molecules coming together to form an aggregate to minimize unfavorable interactions, in this case, polymer contracts into a micelle-like globule. In a poor solvent, the polymer characteristics resemble that of the bulk polymer and become independent of the solvent.

In some polymer–solvent pairs, neither the solvent–polymer nor the polymer–polymer interactions are more favorable than the other. In this case, the interactions may be thought of as cancelling each other, resulting in an ideal solution. Different polymers will have a number of solvents where this condition is satisfied. This condition is called the theta condition and the theta condition is satisfied at a certain temperature, called the theta (θ) temperature or theta point. A solvent at this temperature is called a theta solvent.

The overlap concentration decreases when molecular weight increases. Often, when molecular weight increases, long-chain branches may increase. The high-molecular-weight polymers with long-chain branches occupy a greater hydrodynamic volume per unit mass than the lower molecular-weight polymer.

Viscosity

Viscosity of solutions depends on concentration. When the concentration is very low, the solution viscosity is not much different than pure solvent viscosity. Even at low concentrations, the viscosity of polymer solutions is high due to large differences in size between the polymer and solvent molecules (Teraoka, 2002). Linear and substantially linear polymers behave in a qualitatively predictable way with respect to the relationship of their viscosity to their structure and conformation.

In the study of dilute solutions, it is assumed that each molecule is isolated from all other molecules in the system. In reality, this ideal condition is only an approximation since the interaction between molecules cannot be completely eliminated. Therefore, all data must be extrapolated to infinite dilution. In dilute solutions, the relationship between their viscosity and their structure and conformation depends effectively on the hydrodynamic volume, the volume "swept out" by the molecules as they tumble in solution. At these low concentrations, where there is effectively no interaction between molecules and they are at their most extended conformation, the viscosity may be very close to that of the solvent.

Intrinsic viscosity, $[\eta]$, is a characteristic quantity of a polymer. Intrinsic viscosity can be determined from the limiting behavior of $(1/c)$ $\ln(\eta/\eta_0)$ as $c \to 0$, where c is the concentration of the polymer (wt%), η the viscosity of the solution, and η_0 the viscosity of the pure solvent:

$$[\eta] = \lim_{c \to 0} \left(\frac{\eta/\eta_0 - 1}{c} \right) \qquad (3.10)$$

A plot of the reduced viscosity $(\eta/\eta_0 - 1)/c$ versus concentration, where c is extrapolated to zero concentration, yields the intrinsic viscosity. Measuring at zero concentration $(c = 0)$ would not be meaningful, but this concept of extrapolating to $c = 0$ is very important in polymer characterization and in thermodynamics generally.

A given chain in any good solvent will have the same intrinsic viscosity $[\eta]$. In that sense, $[\eta]$ is a property *intrinsic* to the chain, but it should be noted that the polymer conformation is also partly determined by the solvent. The intrinsic viscosity of a given chain will also be the same in all its theta solvents. The $[\eta]$ value of a chain in a theta solvent will differ from that in a good solvent. If a polymer is measured in two solvents that induce very different conformations (e.g., a helix in one solvent and a random coil in another), then $[\eta]$ differs dramatically.

The reciprocal of intrinsic viscosity is often used to represent the overlap concentration of a given polymer: $c^* = 1/[\eta]$.

The intrinsic viscosity is also given by Mark–Houwink equation:

$$[\eta] = KM^a \qquad (3.11)$$

where M is the molecular weight, and a and K are constants. log–log plots of $[\eta]$ against molecular weight have the intercept of $\log(K)$ and slope of a. The slope contains information about the shape of the polymer molecules in that particular solvent: for spherical, random coil, and rod, the slopes correspond to 0, 0.5–0.8, and 1.8, respectively. For a theta solvent, a is about 0.5, whereas for a good solvent, this becomes about 0.8.

Calibration of the constants K and a for a particular polymer in a particular solvent at a given temperature allows for the determination of the molecular weight by simple measurement of the concentration dependence of the viscosity. Viscosity nomenclature is given in Table 3.1.

Different methods of measuring viscosity are summarized in Chapter 2.2.

In this experiment, different concentrations of polyvinlypyrrolidone (average molecular weights: 10, 29, and 55 kDa) solutions will be

Table 3.1 Polymer Solution Viscosity

	Units	Equation
Relative viscosity	Dimensionless	$\eta = \eta/\eta_0$
Specific viscosity	Dimensionless	$\eta_{sp} = \eta_r - 1 = (\eta - \eta_0)/\eta_0$
Inherent viscosity	dL g^{-1}	$\eta_{inh} = (\ln \eta_r)/c$
Reduced viscosity	dL g^{-1}	$\eta_{red} = \eta_{sp}/c$
Intrinsic viscosity	dL g^{-1}	$[\eta] = \lim_{c\to 0}(\eta_{sp}/c)$
Kinematic viscosity	cSt	$\eta_k = \eta/\eta$

Note: η_0, solvent viscosity; η, polymer solution viscosity.

measured. From the concentration dependence of viscosity, intrinsic viscosity values for a particular molecular weight will be determined. Using Mark–Houwink equation, the shape parameter, a, will be determined using Equation 3.11. Carrying out the same experiment in a different solvent will result in a different a value, showing the determining role of the solvent in the shape of the polymer in solution and whether a solvent is theta, good, or bad.

Pre-laboratory questions

1. What determines the shape of a polymer in a solvent?
2. How do you expect the viscosity of a polymer to change with the molecular weight of the polymer?

Materials

Polyvinlypyrrolidone, PVP (mol. wt.: 10 kDa (PVP10), 29 kDa (PVP29), and 55 kDa (PVP55))
Water
Ethanol (C_2H_5OH)
Capillary viscometer

Procedure

1. Prepare the following stock solutions: 20 wt% PVP10, 15 wt% PVP 29, and 15 wt% PVP 55 in water and ethanol.
2. Using the stock solutions in water, prepare 2, 5, 10, 15, and 20 wt% 10 k PVP; 2, 5, 10, 15, and 20 wt% 20 PVP 29; and 1, 2, 5, 10, and 15 wt% PVP 55 solutions by diluting with water.
3. Measure the viscosity of all the solutions using a capillary viscosity meter. Using a stopwatch, record the time for the liquid level to drop between the etched lines.

4. Determine the viscosity, η, according to the equation, $\eta = tc$, where t represents time in seconds for the liquid to drop and c is the viscometer constant for a given viscometer. Please note that the manufacturer provides the viscometer constant c. This constant can also be determined by using a solvent of known viscosity such as water as reference. (You can also use a viscometer for these measurements where much less liquid may be required and measurements can be performed automatically and very quickly.)
5. Triplicate the measurements and record the average data.
6. Using the stock solutions in ethanol, prepare 2, 5, 10, 15, and 20 wt% 10 k PVP; 2, 5, 10, 15, and 20 wt% 20 k PVP 29; and 1, 2, 5, 10, and 15 wt% PVP 55 solutions by diluting with ethanol.
7. Repeat steps 3–5 for PVP solutions in ethanol instead of water.

Data and observations

1. Determine the intrinsic viscosity by plotting the reduced viscosity $(\eta/\eta_0 - 1)/c$ versus concentration, where c is extrapolated to zero concentration to yield the intrinsic viscosity (Equation 3.10).
2. Make a log–log plot of intrinsic viscosity versus molecular weight and determine α from the slope (Mark–Houwink relationship (Equation 3.11)).
3. Suggest the conformation of the polymer from a.
4. Repeat steps 1–3 for solutions in ethanol.
5. Compare the a values to determine whether water or ethanol is a better solvent PVP.

Data for water

Molecular weight (kDa)	Viscosity	Intrinsic viscosity	a (Mark–Houwink relationship)	Conformation
10				
30				
55				

Data for ethanol

Molecular weight (kDa)	Viscosity	Intrinsic viscosity	a (Mark–Houwink relationship)	Conformation
10				
30				
55				

Clean-up and disposal

1. Discard the solutions.
2. Clean all the glassware and laboratory station.
3. Wash your hands thoroughly before leaving the laboratory.

Post-laboratory questions

1. How do you expect the *a* to change if this experiment were carried out at a higher temperature?
2. Using the *a* for the polymer–solvent employed from the literature, calculate the molecular weights of the polymers used and discuss the possible reasons of error.

Note for the instructor

It can be quite time-consuming to dissolve the polymer in the solvent when preparing the stock solutions. Combination of stirring and ultra-sonic bath should be used. It is suggested that all dilutions from the stock the solutions are also prepared in advance for the students to carry out the measurements. Still, if there are time constraints, different groups can measure the viscosity of polymer solutions of different molecular weights (or in different solvents) and data can be shared.

In this experiment, any set of polymer with different molecular weights can be used choosing appropriate solvents. The concentration of polymer solutions to be prepared depends on the overlap concentration, and so should be adjusted accordingly. The concentrations should be adjusted such that the flow time of the most concentrated polymer solution should be 2–3 times longer than that of the pure solvent.

References

Hunter, R.J. *Introduction to Modern Colloid Science*. Oxford: Oxford University Press, 1994.

Teraoka, I. *Polymer Solutions: An Introduction to Physical Properties*. New York: Wiley, 2002.

Experiment 6: Adsorption of polyelectrolytes on Silica Nanoparticles

Purpose

The aim of this experiment is to adsorb oppositely charged polyelectro-lytes on silica nanoparticles in a layer-by-layer fashion and measure the surface charge after each layer of polyelectrolyte addition to explore the electrostatic interactions between oppositely charged particles.

Theoretical background

A polymer can be adsorbed on the surface of colloidal particles as a result of Coulombic interactions, dipole–dipole interactions, hydrogen bonding, or van der Waals interactions, or some combination of any or all of these forces (Hunter, 1994). The adsorption of polymer onto the particle surface may take place at certain points and the rest of the polymer may either fold onto the surface or trail out, depending on the affinity between the polymer and the solvent. Example of polyelectroyte adsorption on a substrate is shown in Figure 3.11.

The adsorption of a polymer onto the surface of a colloidal particle may lead to stabilization of the particle against flocculation or oppositely can induce flocculation, depending on the polymer–solvent interaction. Either phenomenon may be desirable for a specific application. For example, stabilization is required if particles are desired to be suspended in solution for extended periods of time, such as in personal or homecare products. However, flocculation of particles induced by polymer bridging is widely used in industry in the treatment of mineral ores and in purification of waste water.

Polyelectrolytes are polymers with ionizable groups. In polar solvents, these groups become charged and counterions are released into the solution. Polystyrene sulfonate, polyacrylic, and polymethacrylic acids and their salts are examples of synthetic polyelectrolytes, whereas DNA or proteins are examples of natural polyelectrolytes (Dobrynin and Rubinstein, 2005).

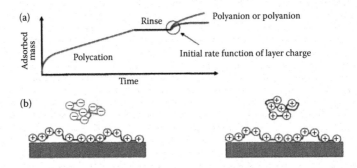

Figure 3.11 Schematic of a multistep experiment, where continuous adsorption of a polycation is halted via a buffer rinse, and followed by adsorption of a (perhaps identical) polycation or a polyanion. (b) Depictions of second step adsorption scenarios: polyanion on an initially adsorbed polycation layer (left) and polycation on an initially adsorbed polycation layer (right). (From Van Tassel, P.R., *Current Opinion in Colloid & Interface Science* 17, 106–113: 2012. With permission.)

As a result of electrostatic interactions between charges, polyelectrolyte solutions behave quite differently from those of uncharged polymers in solution. The crossover from dilute to semidilute solution regime occurs at much lower polymer concentrations for polyelectrolytes. The osmotic pressure of neutral polymers in salt-free solutions exceeds the osmotic pressure of neutral polymers at similar polymer concentrations by several orders of magnitude. The viscosity of polyelectrolyte solutions is proportional to the square root of polymer concentration, whereas viscosity is proportional to polymer concentration for neutral polymers.

Polyelectrolytes adsorbed onto the surface of colloidal particles predominantly through Coulombic interactions. Once a layer of polyelectrolyte is adsorbed onto the particle surface, the surface becomes available for adsorbing oppositely charged polyelectrolytes, again through Coulombic interactions. This leads to a technique known as layer-by-layer adsorption (Figure 3.12). This simple electrostatic assembly technique, with practically no limitations on the shape of charge-bearing species allows for the fabrication of multilayer films from synthetic polyelectrolytes, DNA, proteins, inorganic platelets, nanoparticles, and viruses.

Upon adsorption of polyelectrolytes on colloidal particles, surface charge of the particle changes. In reality, of course, the charge on the particle does not change but the "effective charge" which is the charge that is "seen" by another particle in the same medium changes dramatically. In fact, it is this effective charge that determines the behavior of the particle in suspension, irrespective of whether stabilization or flocculation will occur.

In order to determine the effective charge, the potential at some distance away from the surface of the particle outside the layer of strongly adsorbed ions on the surface needs to be measured. This is done by measuring the electrokinetic or zeta potential (ζ-potential). This measurement is done by putting the suspension in an electric field and measuring the speed of the particles. The principles of this technique are addressed in Chapter 2.3 under electrokinetic techniques.

In this experiment, negatively charged silica nanoparticles will be used as a substrate, and a layer of cationic polyelectrolyte will be adsorbed onto its surface (Dubois et al., 2006). This will be followed by the adsorption of an anionic polyelectrolyte, resulting in a layer-by-layer adsorption. After each layer, the adsorption will be confirmed by the measurement of zeta potential.

Pre-laboratory questions

1. Why coating of a nanoparticle or surface with a polyelectrolyte may be of interest? Find two specific applications.

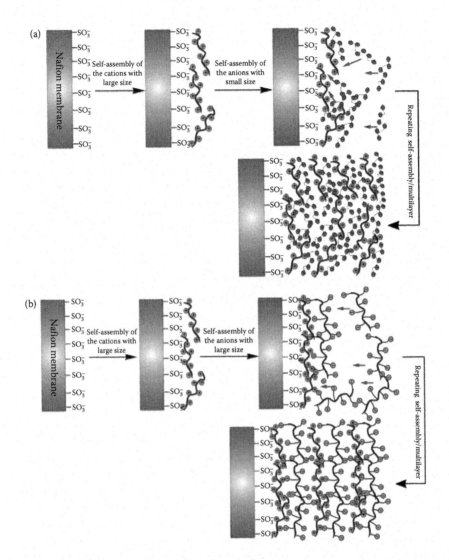

Figure 3.12 Schematic diagram of LbL self-assembly of polycation and polyanion with (a) small-sized monomeric block and (b) with large-sized monomeric block on the Nafion membrane. (From Jiang, S.P. and H. Tang, *Colloids and Surfaces A: Physicochemical and Engineering Aspects* 407, 49–57: 2012. With permission.)

2. Calculate the approximate amounts of PDADMAC, M~100,000–150,000 that would be required to coat 0.5 wt% 100-nm silica nanoparticles and compare this with the suggested concentrations in the experiment. Why do you think exact amounts are not employed?

(Take the contour length of PDADMAC to be 7 Å for one unit and the density of silica to be 2.65 g cm^{-3}.)

Materials

Poly(diallyl dimethylammonium chloride) (PDADMAC, $M \sim 100,000$–150,000)
Poly(allylamine hydrochloride) (PAH, $M \sim 70,000$)
Sodium chloride (NaCl)
100-nm silica dispersion
Conductivity meter
Zeta potential measurement

Procedure

1. Prepare 5 mL of 6 mg mL^{-1} PDADMAC solution (polycation) with 1 mM NaCl.
2. Prepare 5 mL of 6 mg mL^{-1} PAH solution (polyanion) with 1 mM NaCl.
3. Prepare 8 mL of 0.5 wt% salt-free silica colloidal dispersion.
4. Centrifuge the silica dispersion, remove the supernatant (do not discard), and resuspend the particles in water. Using a conductivity meter, measure the conductivity of the supernatant and repeat the centrifugation–decantation–redispersion steps until the conductivity values are less than about 10 µS cm^{-1}.
5. Dilute this solution to obtain 2 mL of 0.1 wt% to measure its zeta potential.
6. Using the remaining 7 mL, add the silica dispersion to 5 mL of the PDADMAC solution dropwise.
7. Leave the suspension under magnetic stirring for 10 min to allow for adsorption to take place.
8. Centrifuge the suspension, remove the supernatant (do not discard), and resuspend the particles in water.
9. Using a conductivity meter, measure the conductivity of the supernatant and repeat the centrifugation–decantation–redispersion steps until the conductivity values are less than about 10 µS cm^{-1} (about three times).
10. Dilute this solution about 10 times to measure the zeta potential of the suspension.
11. Using the PDADMAC-coated silica particle, repeat the same procedure with PAH solution to obtain an anionic colloidal dispersion.
12. After the required centrifugation–decantation–resuspension steps, measure the zeta potential of the obtained dilute suspension.

Data and observation

	Conductivity of decantate 1	Conductivity of decantate 2	Conductivity of decantate 3	Conductivity of decantate 4	Zeta potential
Silica in water					
PDADMAC adsorption					
PAH adsorption					

Clean-up and disposal

1. Collect the solutions in separate collection bottles.
2. Clean all the glassware and laboratory station.
3. Wash your hands thoroughly before leaving the laboratory.

Post-laboratory questions

1. If the conductivity value of the decantate is high, what does this imply as far as its contents?
2. Why is the electrokinetic potential measured as opposed to the measurement of surface charge?
3. How does effective charge of the colloidal particle change upon adsorption of each polyelectrolyte layer?
4. Why is the adsorption of the polycation done first?

Note for the instructor

In this experiment, silica particles of different sizes can also be used. Instead of silica, latex particles can also be coated with polyelectrolytes. Any cationic or anionic polyelectrolyte can be used in this experiment. However, the amount of polyelectrolyte needed to coat the surface of the particles should be adjusted according to the particle used. If an instrument that directly measures the zeta potential is not available, electrophoresis can be used as described in Chapter 2.3. Electrophoresis is the basis of the zeta potential instruments and using an electrophoresis set-up will teach the students more fundamental knowledge about the principles. If electrophoresis is used, two laboratory sessions (2–3 h each) should be allocated for the experiment.

References

Dobrynin, A.V. and M. Rubinstein. Theory of polyelectrolytes in solutions and at surfaces. *Progress in Polymer Science* 30(11), 2005: 1049–118.

Dubois, M., M. Schönhoff, A. Meister, L. Belloni, T. Zemb, and H. Möhwald. Equation of state of colloids coated by polyelectrolyte multilayers. *Physical Review E* 74(5), 2006: 051402.

Hunter, R.J. *Introduction to Modern Colloid Science.* Oxford: Oxford University Press, 1994.

Jiang, S.P. and H. Tang. Methanol crossover reduction by nafion modification via layer-by-layer self-assembly techniques. *Colloids and Surfaces A: Physicochemical and Engineering Aspects* 407, 2012: 49–57.

Van Tassel, P.R. Polyelectrolyte adsorption and layer-by-layer assembly: Electrochemical control. *Current Opinion in Colloid & Interface Science* 17(2), 2012: 106–13.

Experiment 7: Colloidal stability: Flocculation and coagulation

Purpose

The purpose of this experiment is to coagulate electrostatically stabilized colloidal solutions by the addition of different electrolytes and to determine the critical coagulation concentration for each electrolyte. The "salting-out" effect of different anions and cations will be compared.

Theoretical background

The stability of colloidal particles is of paramount importance for the application of colloidal systems. As colloidal particles randomly move in solution as a result of Brownian motion, they collide with each other. During these collisions, if attractive forces dominate, particles stick to each other and remain stuck (Evans and Wennerstrom, 1999).

Most colloidal particles are stabilized by electrostatic repulsion, whereas some particles may be stabilized by steric effects (Hunter, 1994). Some colloidal particles carry a surface charge due to defects in their crystal structure. Others possess surface groups, which can react with acid or base in the medium, resulting in surface charge. Metal oxides often possess an amphoteric metal hydroxide surface layer, which may result in a positive or negative surface charge depending on the pH. When these charged particles are in solution, the oppositely charged counterions are attracted to the surface. At the same time, due to thermal motion, ions have a tendency to dissipate into the bulk. This results in an ion gradient around the particles where some ions are held close to the colloidal surface and the ion concentration gradually decreases

Surface

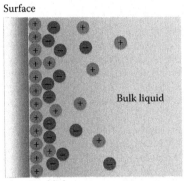

Bulk liquid

Double layer

Figure 3.13 Double layer on a solid surface.

away from the surface. This is the diffuse electrical double layer around the particle. A schematic representation of the double layer is shown in Figure 3.13.

For electrostatically stabilized colloidal systems, a change in the ionic conditions of the medium may lead to the screening of charges, causing the double layer to shrink resulting in particles to flocculate or coagulate. Finally, particles may sediment under gravitational force if the flocculation density is higher than the solution and cream if their density is lower. The minimum amount of electrolyte necessary to induce coagulation is the critical coagulation concentration (ccc). This point corresponds to the point where the maximum in the total potential energy curve touches the horizontal axis. Figure 3.14 shows the total potential energy (V_T) for a stable, metastable, and unstable sol.

The curves corresponding to V_R show different surface potential values but approximately with same concentration (Evans and Wennerstrom, 1999).

At this touch point, the total potential energy and also the derivative of the curve is equal to zero. This concentration strongly depends on the valency of the coagulating ion, which is known as the Schulze–Hardy rule (Schulze, 1882; Hardy, 1899). The rule suggests that the ccc is proportional to the sixth power of the valency of the counterion.

The precipitation of a colloidal suspension upon addition of an electrolyte (salt) is also known as "salting out." This term is more frequently used for polymer and protein solutions. It was discovered that a series of salts have consistent effects on the solubility of proteins. The Hofmeister series (Hofmeister, 1888) is a classification of ions in order of their ability to salt out proteins (Figure 3.15). Anions appear to have a larger effect than cations.

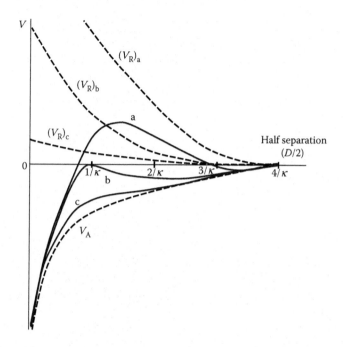

Figure 3.14 The total potential energy of interaction, V_T, for: (a) stable, (b) marginally stable, and (c) unstable sol. (b) The critical coagulation concentration.

Most stabilizing	Most destabilizing
Strongly hydrated anions	Weakly hydrated anions

$$citrate^{3-}>sulfate^{2-}>phosphate^{2-}>F^->Cl^->Br^->I^->NO_3^->ClO_4^-$$

$$N(CH_3)_4^+>NH_4^+>Cs^+>Rb^+>K^+>Na^+>H^+>Ca^{2+}>Mg^{2+}>Al^{3+}$$

Weakly hydrated cations	Strongly hydrated cations

Figure 3.15 The order of cations and anions.

The adsorption of ions from water strongly depends on the hydration energies of different ions. This also has an effect on ion mobility and therefore has a determining effect on the coagulation behavior.

In this experiment, gold nanoparticles will be synthesized and a stable fluid will be obtained (Frens, 1973). The ccc upon addition of salts of different valence will be determined and the "salting out" effect of the different cations and anions will be compared.

Pre-laboratory questions

1. How are Au nanoparticles stabilized in aqueous medium?

2. Once coagulation occurs upon addition of salt, what can you do to obtain stable Au nanoparticles?
3. Can you protect nanoparticles from "salting out"? How?

Materials and equipment

Gold chloride ($HAuCl_4 \cdot 3H_2O$)
Sodium citrate ($Na_3C_6H_5O_7$)
Potassium chloride (KCl)
Barium chloride ($BaCl_2$)
Cerium chloride ($CeCl_3$)

Procedure

Preparation of Au nanoparticles by Frens method (Frens, 1973)
1. Prepare 100 mL of 0.01 wt% $HAuCl_4 \cdot 3H_2O$ solution.
2. Place the solution in a 100-mL Erlenmeyer flask and boil the solution under magnetic stirring.
3. After bubbles are observed, immediately add 2 mL of 1 wt% sodium citrate solution to the boiling solution.
4. After approximately 25 s, the color of the solution turns into blue and then to red.
5. Continue the stirring and boiling for another 5 min.
6. Leave the solution to cool to room temperature.

Determination of critical coagulation concentration

1. Take the Au nanoparticle solution from the synthesis above and measure the pH of the solution.
2. Prepare 50 mL of 0.05 M KCl, $BaCl_2$, and $CeCl_3$ solutions.
3. Divide the gold solution into two and adjust the pH of the solutions to 4 and 7 using an appropriate acid or base.
4. Place 10 mL of 0.01 wt% gold nanoparticles at pH 4 in three test tubes.
5. Add 50 µL of 0.05 M KCl to 10 mL of 0.01 wt% gold nanoparticles at pH 4, shake the solution, and wait for 2 min.
6. Measure the absorbance spectrum between 300 and 800 nm.
7. Continue adding 0.05 M KCl to the gold solution in 50-µL increments until precipitation occurs. This detects the concentration where the gold sol settles to leave a clear supernatant (ccc). After each addition of 50 µL of 0.05 M KCl, measure the absorbance spectrum between 300 and 800 nm.
8. Repeat steps 5–7 for $BaCl_2$ and $CeCl_3$ instead of KCl.
9. Repeat steps 4–8 for 0.01 wt% gold nanoparticles at pH 7.

Data and observations

Total amount	pH 4		pH 7	
KCl (0.05 M)	λ_{max}	Abs_{max}	λ_{max}	Abs_{max}
50 μL				
100 μL				
150 μL				
200 μL				
250 μL				
BaCl$_2$ (0.05 M)	λ_{max}	Abs_{max}	λ_{max}	Abs_{max}
50 μL				
100 μL				
150 μL				
200 μL				
250 μL				
CeCl$_3$ (0.05 M)	λ_{max}	Abs_{max}	λ_{max}	Abs_{max}
50 μL				
100 μL				
150 μL				
200 μL				
250 μL				

Clean-up and disposal

1. Collect the solutions in separate collection bottles.
2. Clean all the glassware and laboratory station.
3. Wash your hands thoroughly before leaving the laboratory.

Post-laboratory questions

1. Why does salt induce coagulation?
2. How does the ccc change with the valence of the salt used?
3. What is the effect of pH on the ccc?
4. How can you correlate the change in the absorbance spectrum with the ccc?

Note for the instructors

Gold synthesis can be carried out as a separate experiment or gold sol can be prepared by the instructor and be provided to the students. The same experiment can be carried out by different groups, for example, using KNO_3, $Ba(NO_3)_2$, and $Ce(NO_3)_3$ to see the similarity of ccc between ions of similar valency. The effect of cation radius can also be investigated by carrying out the same experiment for cations with smaller ionic radii, such as NaCl, $CaCl_2$, and $AlCl_3$, and compare with larger ionic radii, such as K(I), Ba(II), and Ce(III). Different groups can carry out the experiment with a different series of salts and data can be shared to have a better understanding.

The size growth can also be observed with a dynamic light scattering (DLS) method. Once ccc is determined, students can prepare a few gold nanoparticle solutions containing salt below the ccc and measure the size of nanoparticles with DLS as ccc is approached. However, once aggregation starts, the polydispersity increases and it becomes difficult to observe the expected size increase using DLS. After absorbance values between 300 and 800 nm is measured, the shift of maximum absorbance wavelength to higher wavelengths, and lowering of the absorbance at λ_{max} can be detected as ccc is approached, indicating a size growth.

References

Evans, D.F. and H. Wennerstrom. *The Colloidal Domain: Where Physics, Chemistry, Biology, and Technology Meet.* New York: Wiley-VCH, 1999.

Frens, G. Controlled nucleation for regulation of particle-size in monodisperse gold suspensions. *Nature—Physical Science* 241, 1973: 20–22.

Hardy, W.B. A preliminary investigation of the conditions which determine the stability of irreversible hydrosols. *Proceedings of the Royal Society of London* 66(424–433), 1899: 110–25.

Hofmeister, F. Zur Lehre Von Der Wirkung Der Salze (About the Science of the Effect of Salts). *Archiv Fur Experimentelle Pathologie Und Pharmakologie* 24, 1888: 247–60.

Hunter, R.J. *Introduction to Modern Colloid Science.* Oxford: Oxford University Press, 1994.

Schulze, H. Schwefelarsen in Wässriger Lösung. *Journal für Praktische Chemie* 25(1), 1882: 431–52.

Experiment 8: Stabilization of colloids: Emulsions and pickering emulsions

Purpose

The purpose of this experiment is to understand emulsion stabilization by means of solid particles. The relationship between wetting of particles and emulsion stabilization and the factors affecting the emulsion stabilization are explored.

Theoretical background

Emulsion is a general term that is used for mixtures of oil and water in the liquid phase. These two liquid phases are essentially insoluble in each other. As the oil–water interactions are very unfavorable, this kind of a mixture phase separates into two distinct layers. It is possible to reduce the interfacial tension between oil and water by adding a surfactant. Depending on the hydrophobic–lipophilic balance (HLB) of the surfactant, oil-in-water (o/w) or water-in-oil (w/o) emulsions may be formed. It has been shown that the packing parameter of the surfactant at the oil–water interface determines the tendency of the surfactant monolayer to curve toward water or oil or remain effectively planar. Therefore, hydrophilic surfactants result in o/w, whereas more lipophilic surfactants tend to form w/o emulsions. The average particle size of emulsions is in the order of micrometers, which is the upper end of the colloidal size domain. Because of the free energy associated with the oil–water interface, emulsions are thermodynamically unstable and in time phase separate into water and oil layers (Hiemenz and Rajagopalan, 1997; Haase et al., 2010a,b).

Depending on the size of the dispersed phase, emulsions can be classified as emulsions (droplet sizes between 0.1 and 10 μm), miniemulsions (droplet sizes between 100 and 30 nm), and microemulsions (droplet sizes between 30 and 1 nm). Some characteristics of mini- (also known as nano) and micro-emulsions are shown in Table 3.2.

Phase separation between two immiscible liquids minimizes the interfacial area between the two phases. Emulsification results in the creation of a large contact area, thus is energetically unfavorable and requires energy input, usually in the form of vigorous stirring and/or shaking (mechanical energy). The average droplet size is determined by the energy input. Once this external energy is removed, the system has a tendency to revert back to a phase separation, making emulsions unstable. On the other hand, this interfacial tension between immiscible liquids can be lowered upon addition of a surfactant, which in this context can also be called an emulsifier, stabilizing agent, or dispersing agent (chemical energy).

Table 3.2 Characteristics of Emulsions

Type	Color	Stability	Industrial applications
Emulsion	Milky	Unstable	Food, pharmaceuticals, cosmetics
Miniemulsion	Slightly turbid	Kinetically stable	Cosmetics, polymerization, antibacterial
Microemulsion	Transparent	Thermodynamically stable	Recovery of oil, dry cleaning, pesticides

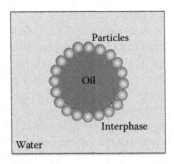

Figure 3.16 Schematic of a single droplet stabilized by solid particles and influence of the contact angle on the preferred emulsion state.

Like all colloidal systems, composition and temperature determine the properties of emulsions but because emulsions are thermodynamically nonequilibrium states, the method with which they are prepared with also play a role in their properties. The resulting emulsion may be o/w or w/o depending on the method of preparation, and therefore for reproducibility, precise definition of preparation procedures is of utmost importance.

In order to obtain emulsions with extended stability, spherical particles, which adsorb at the interface, may be used (Figure 3.16). The fact that finely divided solid particles can act as stabilizer in emulsions is known since the beginning of last century, and therefore solid particles have been present in emulsion formulations in various industries for many years. The credit is usually given to Pickering (1907), who noted that the particles, which were wetted more by water than by oil, act as emulsifiers for o/w emulsions by residing at the interface.

Depending on the contact angle θ_{ow}, which the particle makes with the interface, either one or the other type of emulsion can be formed. Usually, if the $\theta_{ow} < 90°$, particles have a tendency to reside in water and result in the formation of o/w emulsions, whereas $\theta_{ow} > 90°$ leads to the formation of w/o emulsions as a result of the tendency of particles to reside more in oil. In general, in an emulsion containing solid particles, one of the liquids wet the solid more than the other liquid and the more poorly wetting liquid becomes the dispersed phase. However, it should be noted that if the particles are either too hydrophilic (too low θ_{ow}) or too hydrophobic (too high θ_{ow}), they tend to remain dispersed in either the aqueous or oil phase, respectively, giving rise to very unstable emulsions.

In this experiment, silica nanoparticles (~100 nm) will be used to stabilize o/w emulsions. The effect of particle concentration and the polarity of the oil used on the stability of the emulsions will be explored by observing phase separation over time (Aveyard et al., 2003; Binks and Whitby, 2005).

Pre-laboratory questions

1. Why are emulsions thermodynamically unstable?
2. If the contact angle the particle makes with the interface is too high, can a stable emulsion be formed? How?

Apparatus and chemicals

Mechanical stirrer or homogenizer
Light microscope
Cyclohexane (C_6H_{12})
Decane ($C_{10}H_{22}$)
Hexadecane ($CH_3(CH_2)_{14}CH_3$)
Silica particles of 100 nm

Procedure

Emulsions without particles

1. Mix cyclohexane and water in a 1:5 ratio for 10 min using either a homogenizer or a mechanical stirrer ensuring complete mixing.
2. Determine the size of emulsion droplets using a light microscope.
3. Determine the stability of the emulsion by monitoring the emulsion appearance over time.

Effect of particle concentration on the size and stability of emulsion droplets

1. Mix 0.2, 0.5, 1, and 2 wt% silica with a 1:5 mixture of cyclohexane and water for 10 min using either a homogenizer or a mechanical stirrer ensuring complete mixing.
2. Determine the size of emulsion droplets using a light microscope.
3. Determine the stability of the emulsions by monitoring the emulsion appearance over time.

Effect of the polarity of oil on the stability of emulsions with particles

1. Mix 1 wt% silica with a 1:5 mixture of decane and water for 10 min using either a homogenizer or a mechanical stirrer ensuring complete mixing.
2. Mix 2 wt% silica with a 1:5 mixture of hexadecane for 10 min using either a homogenizer or a mechanical stirrer ensuring complete mixing.
3. Determine the size of emulsion droplets using a light microscope.
4. Determine the stability of the emulsions by monitoring the emulsion appearance over time.

Observations and data

		Size of emulsion droplet	Emulsion appearance over time
Particle conc.	0 wt%		
	0.2 wt%		
	0.5 wt%		
	1 wt%		
	2 wt%		
Polarity of oil	Cyclohexane		
	Decane		
	Hexadecane		

Clean-up and disposal

1. Collect the solutions in separate collection bottles.
2. Clean all the glassware and laboratory station.
3. Wash your hands thoroughly before leaving the laboratory.

Post-laboratory questions

1. How does stability of an emulsion change with addition of silica? How is the concentration of silica important?
2. What can be the advantages of using nanoparticles to stabilize emulsions instead of using surfactants as stabilizers?

Note for the instructor

The size of the emulsion droplets that will form strongly depends on the mechanical agitation applied. If mechanical stirring or homogenizer is used, same stirring speeds should be employed for each set of experiment and although the compositions will be different, each experiment initially will give a similar droplet size when the same mechanical agitation is used. In this experiment, the effect of mechanical agitation on the size of emulsion droplets can be dramatically shown by magnetically stirring one solution and using a homogenizer for the same solution and observing the emulsions under a light microscope.

References

Aveyard, R., B.P. Binks, and J.H. Clint. Emulsions stabilised solely by colloidal particles. *Advances in Colloid and Interface Science* 100–102, 2003: 503–46.

Binks, B.P. and C.P. Whitby. Nanoparticle silica-stabilised oil-in-water emulsions: Improving emulsion stability. *Colloids and Surfaces A: Physicochemical and Engineering Aspects* 253 (1–3), 2005: 105–15.

Haase, M.F., D. Grigoriev, H. Moehwald, B. Tiersch, and D.G. Shchukin. Encapsulation of amphoteric substances in a Ph-sensitive pickering emulsion. *The Journal of Physical Chemistry C* 114(41), 2010a: 17304–10.

Haase, M.F., D. Grigoriev, H. Moehwald, B. Tiersch, and D.G. Shchukin. Nanoparticle modification by weak polyelectrolytes for Ph-sensitive pickering emulsions. *Langmuir* 27(1), 2010b: 74–82.

Hiemenz, P.C. and R. Rajagopalan. *Principles of Colloid and Surface Chemistry*, 3rd Edition. New York: Marcel Dekker, 1997.

Pickering, S.U. Cxcvi. emulsions. *Journal of the Chemical Society Transactions* 91, 1907: 2001–21.

Experiment 9: Foam stability

Purpose

The purpose of this experiment is to measure the foam stability of different surfactant solutions by using shake flask method and to understand the concentration and ionic strength on the foam stability of an ionic surfactant.

Theoretical background

Foam is a dispersion of gas bubbles in liquid or in a solid network. Gas bubbles trapped in polymers are valuable light materials widely used in industry (Shaw, 1992). Here, we will focus on dispersion of gas bubbles in liquid. Foams, if formed from pure liquids, are very short-lived, and stable foams can only form in the presence of a surface-active agent. These can also be referred to as foaming agents. For a pure liquid, when a bubble reaches the surface, it immediately breaks and joins other gas molecules above the liquid. When there are surfactants in the medium, surfactants reside at the surface of the bubble and on the surface of the liquid, and the liquid film is stable enough to rise above with the bubble, creating foam (Figure 3.17).

Foam formation increases the interfacial area and increases the free energy of the system. Owing to their large interfacial area, which contributes to the increase in the free energy of the system, foams are not thermodynamically stable but they can remain stable for extended periods of time. Foam destabilizes by bubble collapse (Exerowa and Kruglyakov, 1997). This can either happen because there is a tendency for the liquid films to drain as a result of capillary pressure at the meniscus of the film and become thinner or there is evaporation or gas diffusion through the liquid films. Liquid drains to reach a certain film thickness and may remain stable if there are no disturbing factors such as vibration, dust,

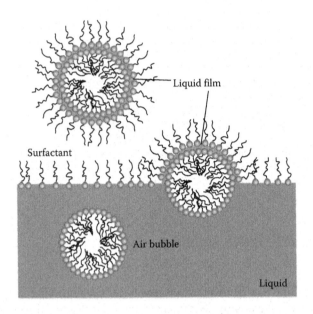

Figure 3.17 Schematic of foam in the presence of surfactants.

evaporation, etc. When the film thickness is below a certain size (200 nm), the thinning process is controlled by surface forces. Attractive forces result in the thinning of the film, whereas repulsive forces lead to a certain thickness, which may result in the so-called stable foam (Waltermo et al., 1996).

When two air bubbles interact, the disjoining pressure that arises due to van der Waals forces is negative, whereas a positive disjoining pressure results due to electrostatic forces. However, these two forces are not solely responsible for the foam stability. Research has shown steric forces, hydration force, and hydrophobic force to also play a role in foam stability (Ruckenstein and Bhakta, 1996; Sedev et al., 1999; Angarska et al., 2004).

Foam stability depends on the rate of lamellae thinning and subsequent bubble rupture. Factors that affect the foam stability are the viscosity of the surface film, surface elasticity, and surface mobility. Temperature, pH, ionic strength, concentration of the surfactant, and surface tension are all factors that affect the foam stability.

The Marangoni effect (or the Gibbs–Marangoni effect) is the mass transfer along an interface between two fluids due to surface tension gradient. Foams stabilize by this effect when surface layers in the lamellae rapidly compensate for surface tension gradient that devolve from the disturbances as foam ages.

Bubbles of a concentrated foam has polyhedral shape. Optimal structure of the foam where the total film area is minimized is a polyhedra of 14 sides, or 12 sides, as have been shown experimentally (Figure 3.18).

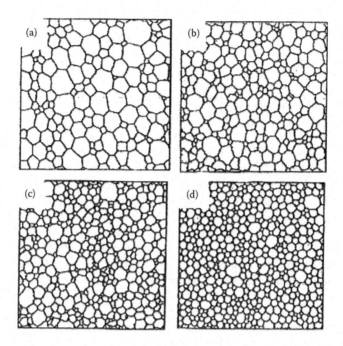

Figure 3.18 Foam structures at different heights: (a) 3 cm, (b) 5 cm, (c) 7 cm, and (d) 9 cm. Transitions from Kugelschaum (spherical) to Polyederschaum (polyhedral) bubbles with increasing froth height. (From Pugh, R.J., *Advances in Colloid and Interface Science*, 114–115, 239–251: 2005. With permission.).

In this experiment, a cationic surfactant, CTAB, will be used to obtain foams (Wang and Yoon, 2004). The effect of surfactant concentration and salinity on foam stability will be observed by measuring the amount of time required for foams to destabilize.

Materials

Graduated cylinder with stopper
Stopwatch
Sodium chloride (NaCl)
Cetyl trimethylammonium bromide (CTAB)

Procedure

1. Prepare 50 mL of 5×10^{-3} M CTAB stock solution.
2. Prepare 50 mL of 5×10^{-2} M NaCl solution.
3. Dilute from the CTAB stock solution to obtain 25 mL of 5×10^{-5}, 1×10^{-4}, 1.5×10^{-4}, CTAB solutions in 50-mL graduated cylinders.

4. Using both the stock solutions, prepare 25 mL of 5×10^{-5}, 1×10^{-4}, 1.5×10^{-4} CTAB solutions with 1×10^{-4} M NaCl in 50-mL graduated cylinders.

5. Similarly, using both the stock solution, prepare 25 mL of 5×10^{-5}, 1×10^{-4}, 1.5×10^{-4} CTAB solutions with 2×10^{-4} M NaCl in 50 mL graduated cylinders.

6. Similarly, using both the stock solutions, prepare 25 mL of 5×10^{-5}, 1×10^{-4}, 1.5×10^{-4} CTAB solutions with 3×10^{-4} M NaCl in 50-mL graduated cylinders.

7. Place stoppers on each graduated cylinder. Repeat the following for all solutions prepared in graduated cylinders one at a time.

8. Flip the graduated cylinder upside down and back for five times at a frequency of 1/2 turn per second (up to down is considered 1/2 a turn).

9. Place the graduated cylinder on a flat surface and start the stop-watch.

10. Record the time (t_1) it takes for a clear liquid surface to form at the center of the foam. Repeat step 7 five times and record the time each time. Take the average of these five readings. It is very important to make sure that there is no foam on the surface of the solution between the readings of the same solution.

11. Repeat the experiment with all surfactant solutions.

12. Plot t_1 (average) in seconds with respect to CTAB concentration.

13. Plot t_1 (average) in seconds with respect to NaCl concentration for each surfactant concentration.

Data and observations

CTAB (M)	NaCl (M)	τ_1
5×10^{-5}		
1×10^{-4}		
1.5×10^{-4}		
5×10^{-5}	1×10^{-4}	
1×10^{-4}	2×10^{-4}	
1.5×10^{-4}	3×10^{-4}	
5×10^{-5}	1×10^{-4}	
1×10^{-4}	2×10^{-4}	
1.5×10^{-4}	3×10^{-4}	
5×10^{-5}	1×10^{-4}	
1×10^{-4}	2×10^{-4}	
1.5×10^{-4}	3×10^{-4}	

Clean-up and disposal

1. Collect the solutions in separate collection bottles.

2. Clean all the glassware and laboratory station.
3. Wash your hands thoroughly before leaving the laboratory.

Post-laboratory questions

1. How does foam stability change with respect to surfactant concentration? Can you explain the trend in terms of surface forces?
2. How does foam stability change with respect to salt concentration in the medium? Can you explain the trend in terms of surface forces?

Note for the instructor

CTAB has a cloud point near the room temperature and if the room is cold, some crystallization may occur. Gentle heating of the solution may be required. When observing the foam stability, it is important that shaking of the solutions is performed in a similar fashion. Shaking them by hand inevitably brings in random errors. To minimize this problem, fixed rotating shakers which flip the flask at a certain frequency and speed can be used. Any surfactant solution can be used for this experiment. However, the concentrations should be adjusted if a different surfactant is to be employed as surfactants form stable foams at high concentrations and foams may not destabilize in an appropriate time for the laboratory session.

References

Angarska, J.K., B.S. Dimitrova, K.D. Danov, P.A. Kralchevsky, K.P. Ananthapadmanabhan, and A. Lips. Detection of the hydrophobic surface force in foam films by measurements of the critical thickness of the film rupture. *Langmuir* 20(5), 2004: 1799–806.

Exerowa, D. and P.M. Kruglyakov. *Foam and Foam Films: Theory, Experiment, Application.* Amsterdam: Elsevier Science, 1997.

Pugh, R.J. Experimental techniques for studying the structure of foams and froths. *Advances in Colloid and Interface Science* 114–115, 2005: 239–51.

Ruckenstein, E. and A. Bhakta. Effect of surfactant and salt concentrations on the drainage and collapse of foams involving ionic surfactants. *Langmuir* 12(17), 1996: 4134–44.

Sedev, R., Z.S. Németh, R. Ivanova, and D. Exerowa. Surface force measurement in foam films from mixtures of protein and polymeric surfactants. *Colloids and Surfaces A: Physicochemical and Engineering Aspects* 149(1–3), 1999: 141–44.

Shaw, D. *Introduction to Colloid and Surface Chemistry*, 4th Edition. London: Butterworth-Heinemann, 1992.

Waltermo, Å., P.M. Claesson, S. Simonsson, E. Manev, I. Johansson, and V. Bergeron. Foam and thin-liquid-film studies of alkyl glucoside systems. *Langmuir* 12(22), 1996: 5271–78.

Wang, L. and R. Yoon. Hydrophobic forces in the foam films stabilized by sodium dodecyl sulfate: Effect of electrolyte. *Langmuir* 20(26), 2004: 11457–64.

Experiment 10: Preparation of colloidal structures using phase diagrams (micelles, liposomes, and microemulsions)

Purpose

To understand the physical differences in a range of colloidal structures and to prepare different colloidal structures from surfactants using binary and ternary phase diagrams.

Theoretical background

Surfactants are surface-active agents with an amphiphilic character; hence the surfactant molecule possesses both hydrophobic and hydrophilic parts. When in a hydrophilic environment, the presence of hydrophobic parts (and when they are in a hydrophobic environment, the presence of hydrophillic parts) leads to self-assembly of the surfactant molecules into a wide range of structures. Depending on the type of surfactant, its concentration, and temperature, surfactants form different phases. These phases are briefly described in Figure 3.19.

Micellar phase: Spherical micelles with an interior composed of hydrocarbon chains and a surface composed of polar headgroups facing water.

Micellar cubic phase: This phase is built up of regular packing of small micelles, which have similar properties to small micelles in the solution phase. The micelles distort to become short prolates rather than spheres, since this allows for a better packing. This phase is highly viscous.

Hexagonal phase: This phase is built up of (infinitely) long cylindrical micelles arranged in a hexagonal pattern, with each micelle being surrounded by six others. The radius of the circular cross-section (which may be somewhat deformed) is close to that of the surfactant molecule length.

Lamellar phase: This phase is built up of bilayers of surfactant molecules with alternating water layers. The thickness of the bilayer is

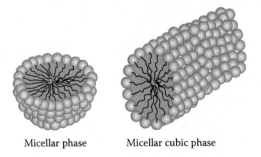

Micellar phase   Micellar cubic phase

Figure 3.19 Different phases of micelles.

somewhat less than twice the surfactant molecule length. The thickness of the water layer can vary (depending on the surfactant) over wide ranges. The surfactant bilayer can range from being stiff and planar to being very flexible and undulating.

Vesicle phase: Vesicles are built from bilayers similar to those of the lamellar phase and are characterized by two distinct water compartments, with one forming the core and the other external medium.

Bicontinuous cubic phase: There can be a number of different structures, where the surfactant molecules form aggregates that penetrate space, forming a porous connected structure in three dimensions (Laughlin, 1996; Holmberg et al., 2002). The amphiphilic interfaces of bicontinuous cubic phase form triply periodic surfaces, which divide space into two unconnected but intertwined labyrinths. Both labyrinths percolate space and provide the pathways which can be used to traverse the structure both in the hydrophobic and hydrophilic regions. In most cases, the two labyrinths are congruent to each other (balanced structure).

For the type of surfactant, there is a generalization called the critical packing parameter (CPP) (Figure 3.20), which relates the headgroup's area (a_0), the extended length (l_{max}), and the volume of the hydrophobic molecule (V) into a dimensionless number, $CPP = (V/l_{max}a_0)$ (Israelachvili et al., 1976). It should be noted that a_0 is not necessarily the physical size of the atoms comprising the headgroup, especially in the case of ionic surfactants; instead, it is the average area occupied per headgroup at the interface/surface, which depends mainly on charge repulsion.

This is often used to predict the type of self-assembly structures that form in solution. Surfactants with a CPP of 1/2 often results in spherical micelles, whereas a CPP is 1/3 points to rod-like micelles. Table 3.3 summarizes the relationship between the structure and CPP.

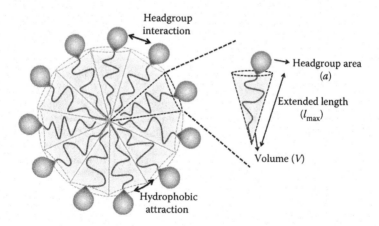

Figure 3.20 Critical packing parameter.

Table 3.3 Classification of Surfactant Aggregates

CPP	Name of the aggregate	Classification I	Classification II	Notes
<1/3	Micellar	Discrete	Isotropic	Micelle is a term borrowed from biology and popularized by G.S. Hartley in his classic book "Paraffin Chain Salts, A Study in Micelle Formation" (Hartley, 1936)
	Cubic micellar	Discrete	Isotropic, liquid crystalline	Highly viscous compared to micellar solutions
1/3–1/2	Hexagonal	1D continuous	Anisotropic, liquid crystalline	Observed at high surfactant or lipid concentrations, birefringent
≈1	Lamellar	2D continuous	Aniosotropic, liquid crystalline	This is an ordered liquid crystalline phase where the surfactant molecules are smectically ordered. Phospholipids have a strong preference to form lamellar phase
	Vesicle	Discrete	Isotropic	Liposomes were first described by British hematologist Dr. Alec D. Bangham at the Babraham Institute, in Cambridge (Bangham and Horne, 1964)
≥1	Bicontinuous cubic	3D continuous	Liquid crystalline	Due to their biological relevance, inverse cubic phases are well investigated with lipid–water mixtures (Luzati et al., 1957; Fontell, 1990; Conn et al., 2006)
	Reversed (inverted) hexagonal	1D continuous	Anisotropic, liquid crystalline	Hexagonal phases may sometimes have alternative arrangements of cylinders or noncircular cross-section of the aggregates
>1	Reversed (inverted) micellar	Discrete	Isotropic	Used as nanoreactors for reactions in aqueous medium (Khomane and Kulkarni, 2008)

These phases can be classified as isotropic and anisotropic (liquid crystalline). Isotropy is uniformity in all directions. Anisotropy is used to describe situations where properties vary systematically, depending on the direction. It can be defined as a difference, when measured along different axes, in a material's physical property (absorbance, refractive index, density, etc.)

Isotropic solutions and anisotropic solutions can be qualified by visual inspection under polarized light. Polarized light is a contrast-enhancing technique that improves the quality of the image obtained with birefringent materials when compared to other techniques. Isotropic materials have only one refractive index, and no restriction on the vibration direction of light passing through them. In contrast, anisotropic materials, which include 90% of all solid substances, have optical properties that vary with the orientation of incident light with the crystallographic axes. They demonstrate a range of refractive indices depending both on the propagation direction of light through the substance and on the vibrational plane coordinates.

Another classification can be discrete or connected. In discrete phases, the oil and water compartments form a closed structure, whereas in connected systems they do not. There may be connectivity in one, two, or three dimensions. Table 3.3 summarizes the classification of surfactant aggregates.

The phase behavior of surfactants is summarized in phase diagrams. Phase diagrams give information about the phases that are present and the composition of each phase. Establishing phase diagrams for a surfactant–H_2O system at different temperatures (binary phase diagrams) or surfactant–H_2O–oil systems (ternary phase diagrams) is possible through extensive investigation of structures at each composition and temperature. Different characterization techniques explained in Chapter 2 are used to shed light on the different self-assembly structures formed. Here are some examples explaining how to find the composition of components at a certain location on the phase diagrams.

Figure 3.21 is a hypothetical phase diagram of a surfactant/H_2O system. There are two possibilities in using the phase diagram. One may know the composition and be interested in finding out what structure forms at that composition. Alternatively, one may be interested in being, for example, in the micellar region and wants to decide on the composition accordingly. Either way, one should draw parallel lines as shown in the figure from the point of interest on the phase diagram to the x- and y-axes which would then correspond to the relevant temperature and composition.

Figure 3.22 is a hypothetical phase diagram of a surfactant–H_2O–oil system. Three components are presented in a triangle. It is a little more complicated to find your way around these phase diagrams. These phase diagrams are at a fixed temperature and if temperature is also a parameter, it enters the phase diagram as the fourth component and is shown as

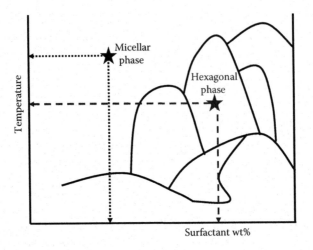

Figure 3.21 Example of a binary phase diagram.

a three-dimensional figure. These phase diagrams are a little more complex and beyond the scope of this book. To determine the composition of a point on the phase diagram, typically one draws three straight lines passing through that point, each line being parallel to one of the axes as shown in Figure 3.23 (Evans and Wennerstrom, 1999; Holmberg et al., 2002). The concentration of *X* is the concentration read from the baseline of *YZ*, the concentration of *Y* is the concentration read from the baseline of *XZ*, and the concentration of *Z* is the concentration read from the baseline of *XY*.

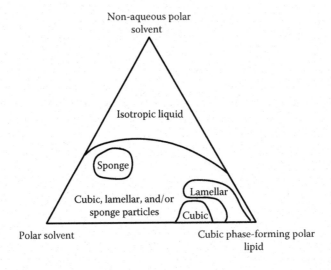

Figure 3.22 Example of a ternary phase diagram.

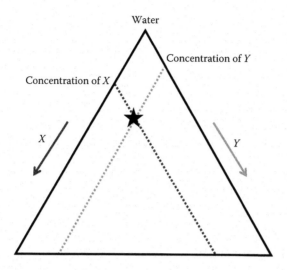

Figure 3.23 Determination of composition from a ternary phase diagram.

Depending on how the concentration for each component is presented (may be pure substance or wt% concentration), the corresponding composition of a certain point on the ternary phase diagram can be found.

In this experiment, different self-assembly structures will be formed by mixing surfactant and water (and oil, when required) at given ratios determined from binary and ternary phase diagrams. Differences in the appearance (turbidity, viscosity, etc.) of these dispersions will be observed and samples will be inspected under polarized light for the detection of anisotropic phases.

Pre-laboratory questions

1. According to your understanding of different colloidal structures, can you have an estimate about some of the physical aspects of each type of structure? Which ones do you expect to be: Clear solution? Turbid solution? Highly viscous? Birefringent?
2. Calculate the critical packing parameter for sodium dodecyl sulfate (SDS) and dodecyltrimethylammonium chloride (DTAC), using required values from the literature.

Materials and chemicals

Automatic pipettes and tips
Test tubes
Dioctyl sodium sulfosuccinate (AOT)
Cetyltrimethylammonium bromide (CTAB)
Dodecyltrimethylammonium chloride (DTAC)

Monopentadecenoin (MPD)
Sodium dodecyl sulfate (SDS)
Sodium octyl sulfate (SOS)

Procedure

1. Preparation of a micellar phase:
   Using the phase diagram 1 (Figure 3.24), prepare a 10-mL micellar solution and note your concentrations.
2. Preparation of a micellar cubic phase:
   Using the phase diagram 2 (Figure 3.25), prepare a 10-mL micellar cubic solution and note your concentrations.
3. Preparation of a hexagonal phase:
   Using the phase diagram 1, prepare a 10-mL hexagonal solution and note your concentrations.
4. Preparation of a lamellar phase:
   Using the phase diagram 1, prepare a 10-mL lamellar solution and note your concentrations.
5. Preparation of a vesicle phase:
   Using the phase diagram 3 (Figure 3.26), prepare a 10-mL vesicle solution and note your concentrations.
6. Preparation of a bicontinuous cubic phase:
   Using the phase diagram 2, prepare a 10-mL bicontinuous cubic solution and note your concentrations.

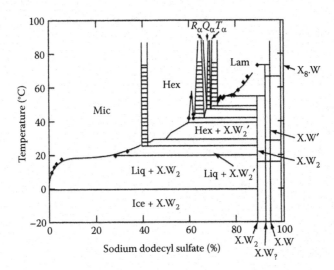

Figure 3.24 Phase diagram 1: Binary phase diagram of sodium dodecyl sulfate (SDS)–water. (From Kékicheff, P. et al., *Journal of Colloid and Interface Science*, 131, 112–132: 1989. With permission.)

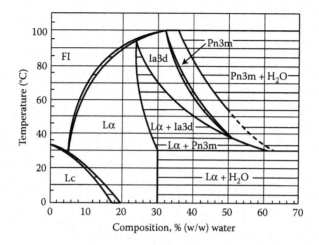

Figure 3.25 Phase diagram 2: Binary phase diagram of monopentadecenoin/water. (From Briggs, J. and M. Caffrey, *Biophysical Journal* 67: 594–1602, 1994. With permission.)

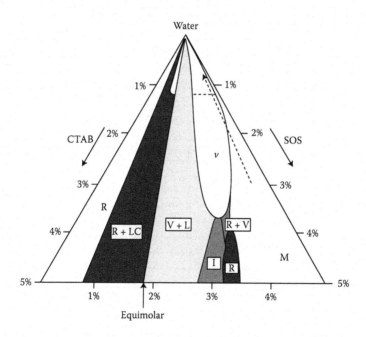

Figure 3.26 Phase diagram 3: Ternary phase diagram of sodium octyl sulfate (SOS)/cetyltrimethylammonium bromide (CTAB)/water. (From Yatcilla, M. et al., *Journal of Physical Chemistry* 100, 5874–5879:1996. With permission.)

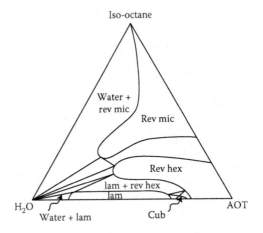

Figure 3.27 Phase diagram 4: Ternary phase diagram of dioctyl sodium sulfos-uccinate (AOT)/water/isooctane. (From Tamamushi, B. and N. Watanabe, *Colloid and Polymer Science* 258, 1980: 174–178. With permission.)

7. Preparation of reversed hexagonal phase:
   Using the phase diagram 4 (Figure 3.27), prepare a 10-mL reversed hexagonal solution and note your concentrations.
8. Preparation of a reversed micellar phase:
   Using the phase diagram 4, prepare a 10-mL reversed micellar solution and note your concentrations.
9. Fill in the table with your observations and calculations.

Observations

Name of the aggregate	CPP	Appearance	Viscosity	Birefringency
Micellar				
Cubic micellar				
Hexagonal				
Lamellar				
Vesicle				
Bicontinuous cubic				
Reversed hexagonal				
Reversed micellar				

Clean-up and disposal

1. Collect the solutions in separate waste containers.
2. Clean all the glassware and laboratory station.
3. Wash your hands thoroughly before leaving the laboratory.

Post-laboratory questions

1. What techniques would you suggest to use to exactly determine the structures in each phase? (Hint: You may need to suggest more than one technique to be used in combination.)
2. Using phase diagram 3, describe how you would obtain micelles from the vesicle solution that you prepared. Describe where this would move you on the phase diagram.
3. Using phase diagram 1, describe how you would obtain micelles from the hexagonal solution that you prepared. Describe where this would move you on the phase diagram.

Note for the instructor

Different phase diagrams of other surfactant systems can be found in the literature to be used in this experiment. When preparing the reversed structures from oil–surfactant–water systems, the surfactant should first be solubilized in oil before the addition of the water phase. For closed-packed systems, a paste rather than a solution may be obtained, so using vials instead of volumetric flasks may be more appropriate.

References

Bangham, A.D. and R.W. Horne. Negative staining of phospholipids and their structural modification by surface-active agents as observed in the electron microscope. *Journal of Molecular Biology* 8(5), 1964: 660–68.

Briggs, J. and M. Caffrey. The temperature-composition phase diagram and mesophase structure characterization of monopentadecenoin in water. *Biophysical Journal* 67, 1994: 1594–602.

Conn, C.E., O. Ces, X. Mulet, S. Finet, R. Winter, J.M. Seddon, and R.H. Templer. Dynamics of structural transformations between lamellar and inverse bicontinuous cubic lyotropic phases. *Physics Review Letters* 96, 2006: 108102.

Evans, D.F. and H. Wennerstrom. *The Colloidal Domain: Where Physics, Chemistry, Biology, and Technology Meet*, New York: Wiley-VCH, 1999.

Fontell, K. Cubic phases in surfactant and surfactant-like lipid systems. *Colloid and Polymer Science* 268, 1990: 264–85.

Hartley, G.S. *Aqueous Solutions of Paraffin-Chain Salts: A Study in Micelle Formation.* Paris: Hermann & Cie, 1936.

Holmberg, K., B. Jonsson, B. Kronberg, and B. Lindman. *Surfactants and Polymers in Aqueous Solution*, 2nd Edition. Chichester: Wiley, 2002.

Israelachvili, J.N., D.J. Mitchell, and B.W. Ninham. Theory of self-assembly of hydrocarbon amphiphiles into micelles and bilayers. *Journal of the Chemical Society, Faraday Transactions 2: Molecular and Chemical Physics* 72, 1976: 1525–68.

Kékicheff, P., C. Grabielle-Madelmont, and M. Ollivon. Phase diagram of sodium dodecyl sulfate-water system: 1. A calorimetric study. *Journal of Colloid and Interface Science* 131(1), 1989: 112–32.

Khomane, R.B. and B.D. Kulkarni. Nanoreactors for nanostructured materials. *International Journal of Chemical Reactor Engineering* 6, 2008: A62.

Laughlin, R.G. *The Aqueous Phase Behavior of Surfactants.* San Diego: Academic Press, 1996.

Luzzati, V., H. Mustacchi, and A. Skoulios. Structure of the liquid-crystal phases of the soap–water system: Middle soap and neat soap. *Nature* 180, 1957: 600–1.

Tamamushi, B. and N. Watanabe. The formation of molecular aggregation structures in ternary system: Aerosol ot/water/iso-octane. *Colloid and Polymer Science* 258, 1980: 174–78.

Yatcilla, M.T., K.L. Herrington, L.L. Brasher, E.W. Kaler, S. Chiruvolu, and J.A. Zasadzinski. Phase behavior of aqueous mixtures of cetyltrimethylammonium bromide (CTAB) and sodium octyl sulfate (SOS). *The Journal of Physical Chemistry* 100(14), 1996: 5874–79.

Experiment 11: Preparation of miniemulsions by phase inversion

Purpose

The purpose of this experiment is to prepare a w/o miniemulsion by using a low energy method, namely phase inversion. The stability of these miniemulsions will be investigated in the absence and presence of an anionic surfactant.

Theoretical background

Emulsions with droplet sizes in the range 20–200 nm form a separate class of emulsions, called miniemulsion. Owing to their small size in the nanometer range, in the past 20 years, nanoemulsion has become an alternative word to describe these systems. These droplet sizes are achieved with the help of a surfactant (stabilizer); however, these systems are not thermodynamically in equilibrium like microemulsions, where smaller droplet sizes are observed. However, they are kinetically stable due to the presence of an adsorbed layer at the oil–water interface, causing an electrostatic or steric barrier against coagulation. As a result of the small size of the droplets in the continuous phase, miniemulsion dispersions appear nearly transparent, definitely translucent. Owing to their small size, Brownian motion results in faster diffusion rates than sedimentation and coagulation, providing long-term stability against flocculation and sedimentation, making them interesting in certain applications, ranging from cosmetics to chemical industries (Capek, 2004). More studies have been done on o/w systems, especially in using them as nanoreactors for polymerization reactions. An example of a polymerization in miniemulsions is the Experiment 16 of this book.

Miniemulsions are not thermodynamically stable systems and therefore they do not spontaneously form. Some energy input is required into the system to obtain miniemulsion. Some methods require high energy such as high-shear stirring, homogenization, ultrasound sonication, etc. The most widely used emulsifying method is the use of high-pressure homogenizers, resulting in small-size miniemulsions in a short time. Another approach is the use of low-energy emulsification methods, such as phase inversion, which takes advantage of the change in the solubility of the stabilizer with temperature during emulsification process (Solans et al., 2005).

Phase inversion

Low-energy phase inversion method was introduced by Shinoda and Saito (1968). Initially an emulsion may contain equal amounts of oil and water phases and an emulsifier suitable for o/w emulsion. A practical method to choose a suitable emulsifier is using the HLB (Griffin, 1954). This is a classification only for nonionic surfactants and is based on the ratio of the hydrophilic part of the molecule to the hydrophobic (lipophilic) part:

$$\text{HLB} = \frac{m_h}{m} \times 20 \qquad (3.12)$$

where m_h is the mass of the hydrophilic part of the molecules and m the mass of the total molecules. This gives a range between 0 and 20, where 0 and 20 correspond to completely hydrophobic and completely hydrophilic molecules, respectively.

To make w/o emulsions, an HLB of 4–6 and for o/w emulsions HLB of 8–16 are suitable. When an o/w emulsion is heated, the hydration (especially when epoxy groups are present) of hydrophobic parts (i.e., tail) decreases and the emulsifier becomes more hydrophobic (corresponds to a lower HLB). At a certain temperature, called the phase inversion temperature, the HLB decreases so much so that the emulsion reverts from an o/w emulsion to an w/o emulsion. As w/o emulsions have much lower conductivities than that of o/w, phase inversion can be followed by a sudden decrease in conductivity as temperature increases.

w/o miniemulsions can be prepared with this method by carefully choosing the PIT to be at least 20 K or more below the storage temperature, so the stability of the emulsion is enhanced while retaining the small droplet size. During cooling, the system crosses a point of zero spontaneous curvature and minimal surface tension, promoting the formation of finely dispersed water droplets (see vertical arrow in Figure 3.28). Rapid cooling of the heated emulsions results in kinetically trapped stable

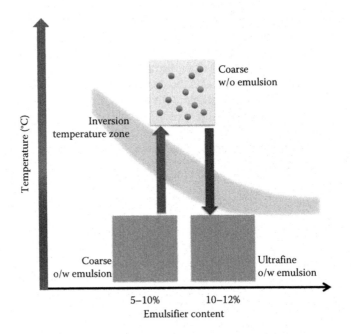

Figure 3.28 Temperature effect on the formation of emulsions.

miniemulsions with small droplet sizes and narrow size distributions (Fernandez, 2004).

In this experiment, an o/w miniemulsion dispersion will be prepared using a nonionic surfactant (BRIJ-L4) with an HLB of 9.7 by phase inversion. The phase inversion temperature of the surfactant will be determined by conductivity measurements. The effect of addition of an anionic surfactant (SDS) on miniemulsion stability will be investigated.

Pre-laboratory questions

1. Are miniemulsions thermodynamically stable? If not, how can they stay stable over a period of time?
2. What happens right at the phase inversion temperature?
3. How does Oswald ripening affect the miniemulsions?

Apparatus and chemicals

Sodium chloride (NaCl)
Hexadecane ($C_{16}H_{34}$) or tetradecane ($CH_3(CH_2)_{12}CH_3$), dodecane ($CH_3(CH_2)_{10}CH_3$)
Polyethylene glycol dodecyl ether (BRIJ 30 or BRIJ-L4)

Sodium dodecyl sulfate (SDS)
Magnetic stirrer
Conductivity meter

Procedure

To prepare 20 mL of o/w miniemulsion:

1. Prepare 100 mL of 10 mM NaCl stock solution.
2. Prepare 50 mM SDS stock solution.
3. Prepare an ice bath.
4. Take 4 mL of hexadecane (oil phase, tetradecane, or dodecane can also be used), 1.2 mL of BRIJ-L4, and 14.8 mL of 10 mM NaCl (aqueous phase) and place them in a beaker with a magnetic stirrer. Addition of NaCl in the aqueous phase is used for the detection of the phase inversion point by measuring conductivity drop during the heating process later.
5. Mix the contents. You will observe a milky solution.
6. Insert a conductivity meter electrode in the solution and measure conductivity as you heat the emulsion.
7. When conductivity drops to zero, quickly place the solution in the ice bath to rapidly cool the solution.
8. Separate this solution into two, of 10 mL each.
9. One part, dilute with 50 mM SDS (anionic surfactant) to obtain a final SDS concentration of 10 mM.
10. The other part, dilute with equal amount of water.
11. Leave the two miniemulsions and observe for a period of 24 h and then daily for 1 week.

Observations and data

$T_{inv} = \ldots\ldots\ldots °C$

Type	2 h	4 h	6 h	8 h	24 h	Day 2	Day 3	Day 4	Day 5	Day 6	Day 7
Miniemulsion											
Miniemulsion + SDS											

Clean-up and disposal

1. Collect the solutions in separate waste containers.
2. Clean all the glassware and laboratory station.
3. Wash your hands thoroughly before leaving the laboratory.

Post-laboratory questions

1. What is the purpose of the addition of NaCl in the medium?
2. Why does measuring conductivity give us the temperature of inversion?
3. How does adding an anionic surfactant change the miniemulsions? Through which mechanism anionic surfactant adds to the stability of the miniemulsion?
4. Do the miniemulsions phase separate?
5. How does turbidity change with time? (One can measure turbidity using a spectrophotometer taking measurements at set time intervals for quantification or just do an eye observation for qualitative observation.)
6. How would you expect the miniemulsion stability to change if a shorter alkane were used as the dispersed phase?

Note for the instructor

Phase inversion temperature is different for each surfactant. Different groups can use a different surfactant and comparisons can be made. The stability of the miniemulsion depends on the oil used. Differences can be observed if different groups perform the experiment using different oil. Any ionic surfactant can be used for the experiment; however, the amount of surfactant required to enhance stability changes, therefore should be adjusted. Although the preparation of the miniemulsion solutions can be done in a single laboratory session, the stability should be observed over a period of days to see significant differences between solutions.

References

Capek, I. Degradation of kinetically-stable O/W emulsions. *Advances in Colloid and Interface Science* 107(2–3), 2004: 125–55.

Fernandez, P., V. André, J. Rieger, and A. Kühnle. Nano-emulsion formation by emulsion phase inversion. *Colloids and Surfaces A: Physicochemical and Engineering Aspects* 251(1–3), 2004: 53–8.

Griffin, W.C. Calculation of Hlb values of non-ionic surfactants. *Journal of the Society of Cosmetic Chemists* 5, 1954: 249–56.

Shaw, D. *Introduction to Colloid and Surface Chemistry*, 4th Edition. Butterworth-Heinemann, 1992.

Shinoda, K. and H. Saito. The effect of temperature on the phase equilibria and the types of dispersions of the ternary system composed of water, cyclohexane, and nonionic surfactant. *Journal of Colloid and Interface Science* 26(1), 1968: 70–4.

Solans, C., P. Izquierdo, J. Nolla, N. Azemar, and M.J. Garcia-Celma. Nano-emulsions. *Current Opinion in Colloid & Interface Science* 10(3–4), 2005: 102–10.

chapter four

Nanoparticles

Brief information about nanofluids

Modern technology allows for the fabrication of materials at the nanometer scale. Nanofluids consist of ultrafine metallic or nonmetallic nanoparticles in a base fluid (polar or nonpolar) where the nanoparticles are generally in the size range 5–60 nm with a surfactant or polymer coating to prevent the agglomeration of nanoparticles. A nanofluid does not mean a simple mixture of solid nanoparticles with a liquid. Special techniques are developed to obtain good dispersion of nanoparticles in liquids or to directly produce stable nanofluids (Murshed et al., 2008).

The size, morphology, and physical properties (thermal, magnetic, etc.) of nanoparticles and their stability in different fluids are determined by the synthesis method, synthesis conditions, and choice of precursors, reducing, as well as stabilizing agents. An unavoidable problem associated with the synthesis of nanosized particles is their instability over longer periods of time. The main reason for the instability of these nanofluids is their tendency to form agglomerates in order to reduce their energy associated with the high-surface-area-to-volume ratio. Moreover, naked metallic nanoparticles are chemically highly active and are easily oxidized in air, resulting generally in loss of dispersity and stability. For many applications, it is thus crucial to develop protection strategies to prevent the agglomeration and obtain stable colloidal solutions. These strategies comprise coating with organic species, including surfactants or polymers, or coating with an inorganic layer, such as silica or carbon. Depending on the desired application, these protecting shells can be used not only as stabilizers, but also as organic and surface bonded or donor ligands (Lu et al., 2007). The nanoparticle fabrication techniques can be classified into two broad categories: top-down and bottom-up (Yu et al., 2008). The "top-down" approach involves the division of solid particles into smaller portions. This approach may involve milling or attrition, chemical methods, and volatilization of a solid followed by condensation of the volatilized components. On the other hand, the "bottom-up" approach, which is far more popular in the synthesis of nanoparticles, involves the condensation of atoms or molecular entities in a gas phase or in solution (Barron, 2009).

The effective usage of nanofluids requires successful synthesis procedures for creating stable suspensions of nanoparticles in liquids. Depending on the requirements of a particular application, many combinations of particle materials and fluids are of potential interest. For example, nanoparticles of oxides, nitrides, metal carbides, and nonmetals with or without surfactant molecules can be dispersed into fluids such as water, ethylene glycol, or oils for the enhancement of heat-transfer properties of these fluids (Keblinski et al., 2005). The use of nanoparticles in optical, mechanical, and electronic devices is dependent on their size and shape. The control over their morphology and size is thus an important goal. Specific control of shape and size is often difficult. Generally, this is achieved by varying the synthesis method, reducing agent, stabilizer, pH, and temperature of the reaction system (Cheon and Park, 2012).

Experiments in this chapter are designed to give a glimpse of different methods of synthesizing nanoparticles as well as making them stable. Metal, metal oxide, and organic nanoparticles will be synthesized and principles of nanoparticle formation will be explored in this chapter.

References

Barron, A. Introduction to Nanoparticle Synthesis. *Connextions*. http://cnx.org/content/m22372/1.2/.

Cheon, J.Y. and W.H. Park. Study on Synthesis of PVA Stabilized Silver Nanoparticles Using Green Synthesis and Their Application for Catalysis. Paper Presented at the Materials Research Society Proceedings, San Francisco, 2012.

Keblinski, P., J.A. Eastman, and D.G. Cahill. Nanofluids for Thermal Transport. *Materials Today* 8(6), 2005: 36–44.

Lu, A.-H., E.L. Salabas, and F. Schuth. Magnetic Nanoparticles: Synthesis, Protection, Functionalization, and Applications. *Angewandte Chemie* 46, 2007: 1222–44.

Murshed, S.M.S., K.C. Leong, and C. Yang. Thermophysical and Electrokinetic Properties of Nanofluids—A Critical Review. *Applied Thermal Engineering* 28, 2008: 2109–25.

Yu, W., D.M. France, J.L. Routbort, and S.U.S. Choi. Review and Comparison of Nanofluid Thermal Conductivity and Heat Transfer Enhancements. *Heat Transfer Engineering* 29(5), 2008: 432–60.

Experiment 12: Synthesis of metal (silver) nanoparticles (reduction)

Purpose

The purpose of this experiment is to obtain silver (Ag) nanoparticles through reduction reaction using a reducing agent in the presence of a dispersing agent.

Theoretical background

Silver is one of the oldest known metals to mankind and it is long been valued as a precious metal. It is a soft, lustrous transition metal, which has the highest electrical conductivity and the highest thermal conductivity than any other metals. Today, silver metal is used in electrical contacts and conductors, in mirrors, and in catalysis of chemical reactions.

Silver nanoparticles are silver particles of sizes <100 nm. Because of their size, these particles have unique properties that are very different from bulk silver metal. Silver nanoparticles exhibit special optical, electrical, and magnetic properties, which make them valuable in several applications ranging from catalysis to biomedical. Most widespread applications utilize the antibacterial and antifungal properties of silver nanoparticles. The size and shape have been shown to throw an impact on the efficacy of particles' antimicrobial properties.

On the other hand, exposure to silver nanoparticles has been associated with "inflammatory, oxidative, genotoxic, and cytotoxic consequences"; the silver particulates have been shown to primarily accumulate in the liver but have also been shown to be toxic in other organs including the brain (Ahamed et al., 2010; Johnston et al., 2010).

There are many different synthetic routes to produce silver nanoparticles. Amongst them, wet chemistry is of interest to us. Wet chemistry synthesis routes use the principle of reduction of a silver salt with a reducing agent in the presence of a colloidal stabilizer. This is a method frequently used to reduce other metals such as gold, copper, nickel, and zinc to obtain nanoparticles. Sodium borohydride is a common reducing agent. Stabilizers can be surfactants or synthetic and natural polymers. Citrate and cellulose have been used both as reducing and as stabilizing agents.

Over the past decades, green chemistry methods have been applied to design greener nanomaterials by reducing the use of hazardous chemicals. A convenient method to achieve stable colloidal solutions of silver nanoparticles having a narrow size distribution using green chemistry methods has been reported (Medina-Ramirez et al., 2009). The synthesis is carried out at low temperatures using green reducing agents.

In this experiment, hydrophilic silver nanoparticles are obtained through the reaction of silver nitrate ($AgNO_3$) with sodium borohydride ($NaBH_4$), where Gum Arabic serves as a dispersing agent.

The procedure is primarily based on a reduction reaction: the chemical reaction is the reduction of silver nitrate in the presence of sodium borohydride as shown below:

$$AgNO_3(aq) + NaBH_4(aq) \rightarrow Ag(s) + \tfrac{1}{2}H_2(g) + \tfrac{1}{2}B_2H_6(g) + NaNO_3(aq)$$

The borohydride anions adsorb onto silver nanoparticles and the repelling forces of the borohydride anions prevent the aggregation of particles (Figure 4.1). However, this stabilization can be hindered by

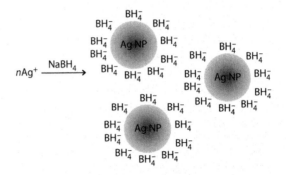

Figure 4.1 Schematic representation of the silver nanoparticle synthesis reaction.

the addition of an electrolyte (see Experiment 7: Colloidal Stability: Flocculation and Coagulation), and therefore a stabilizer (e.g., Gum Arabic) is often used to prevent agglomeration.

After the reduction of silver ions, the concentration of silver atoms increases until supersaturation, after which aggregation will occur. The size of the initial particles depends on the number of nucleation sites. The more the nucleation sites, the smaller will be the particles. The amount of nucleation sites depends on the amount of reducing agent. If there is too little reducing agent, then there will be only few nucleation sites. On the other hand, if there is too much of the reducing agent, then the nucleation sites will form at different times and a wide size distribution of particles will be obtained. Therefore, the amount of ions and reducing agent should be optimized to obtain the desired size and size distribution.

Pre-laboratory questions

1. What type of silver nanoparticles will be obtained through this synthesis method? Hydrophilic or hydrophobic?
2. Find some present applications of silver nanoparticles as antibacterial agents and explain the importance of size for these applications.
3. What is the structure of Gum Arabic? Which part of the molecule do you expect to have an affinity to the Ag surface?

Materials

Silver nitrate ($AgNO_3$)
Sodium borohydrate ($NaBH_4$)
Distilled water (H_2O)
Gum Arabic

Procedure

1. Soak all the glassware in chromic acid and leave overnight. Then rinse them with distilled water.
2. Prepare 0.1 M $AgNO_3$, 0.1 M $NaBH_4$, and 3 wt% Gum Arabic solutions. $AgNO_3$ and $NaBH_4$ solutions should be prepared freshly before starting the experiment.
3. Stir (using a magnetic stirrer) aqueous Gum Arabic solution (3% (m/v) 50 mL) in a round-bottom flask for 30 min at 60°C.
4. Add 3.5 mL of 0.1 M $AgNO_3$ solution to it under continuous stirring and keep the mixture at these conditions for 15 min to allow for diffusion of metal ions to the pores of the dispersing agent.
5. Inject 3.5 mL of reducing agent 0.1 M $NaBH_4$ into the solution incrementally under continuous agitation.
6. Heat the mixture to 85°C and stir at that temperature for 3 h for the hydrolysis of excess dispersing agent, which results in the formation of silver nanoparticles.
7. Precipitate the obtained colloidal suspension of silver nanoparticles with absolute ethanol and centrifuge the dispersion at 6000 rpm for 30 min.
8. Remove the supernatant and repeat the centrifugation two or three times according to the samples obtained.
9. Dry the nanoparticles in a vacuum oven overnight.
10. Suspend the particles in distilled water. Use a spectrophotometer to obtain the absorbance spectrum of the particles between 350 and 800 nm.
11. Measure the size of the silver nanoparticles using dynamic light scattering (DLS).

Data and observations

1. What is the color of the silver nanoparticle solution?
2. Were the dried particles readily suspended in distilled water or was sonication required?
3. Using your spectrophotometer data, from the position of the wavelength that corresponds to the maximum absorbance, estimate the size of the particles using Figure 4.2.
4. Express your results from DLS, in terms of mean size and size distribution and compare with spectrophotometric estimation of size.

Clean-up and disposal

1. Collect the solutions in separate waste containers.
2. Clean all the glassware and laboratory station.
3. Wash your hands thoroughly before leaving the laboratory.

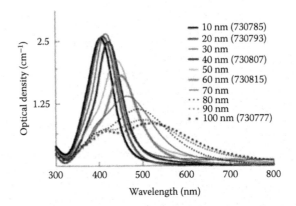

Figure 4.2 The extinction spectra of silver nanoparticles with diameters ranging from 10 to 100 nm. (From Oldenburg, S. J. http://www.nanocomposix.com, 2012. With permission.)

Post-laboratory questions

1. What is the role of Gum Arabic in this experiment?
2. What do you think is the reason for freshly preparing silver nitrate and sodium borohydride solutions for the experiment?
3. Can you suggest alternative reducing agents or dispersing agents?

Note for the instructor

Citric acid can be used instead of Gum Arabic. The principle of this experiment where reduction of a metal ion to obtain nanoparticles is similar to Au nanoparticle synthesis is described in Experiment 7. If Au nanoparticle synthesis is done by the students for Experiment 7, this experiment becomes redundant.

References

Ahamed, M., M.S. AlSalhi, and M.K.J. Siddiqui. Silver Nanoparticle Applications and Human Health. *Clinica Chimica Acta* 411(23–24), 2010: 1841–48.

Johnston, H.J., G. Hutchison, F.M. Christensen, S. Peters, S. Hankin, and V. Stone. A Review of the *In Vivo* and *In Vitro* Toxicity of Silver and Gold Particulates: Particle Attributes and Biological Mechanisms Responsible for the Observed Toxicity. *Critical Reviews in Toxicology* 40(4), 2010: 328–46.

Medina-Ramirez, I., S. Bashir, Z.P. Luo, and J.L. Liu. Green Synthesis and Characterization of Polymer-Stabilized Silver Nanoparticles. *Colloids and Surfaces B-Biointerfaces* 73, 2009: 185–91.

Oldenburg, S. J. Silver Nanoparticles: Properties and Applications. Nanocomposix, http://www.nanocomposix.com, 2012.

Experiment 13: Synthesis of magnetic nanoparticles

Purpose

The purpose of this experiment is to carry out coprecipitation method to obtain Fe_3O_4 magnetic nanoparticles with and without a surfactant coating and explore each step to develop an understanding of the factors affecting the particle formation and stabilization.

Theoretical background

Iron oxides are naturally occurring compounds that are commonly found and are easily synthesized. Different iron oxides exist in the form of oxides, hydroxides, or oxide-hydroxides. Almost all of these materials are crystalline. Information on common iron oxides is given in Table 4.1.

Magnetic properties

Iron oxides have attracted attention because of their magnetic properties. Magnetic properties of solid materials are described along with their magnetic susceptibility (χ), permeability (μ), and magnetic moment (M). Magnetic susceptibility, χ is the magnetic response of a material to an applied magnetic field. The tendency of the magnetic lines to pass through a medium rather than passing through vacuum is the magnetic permeability μ and is related to magnetic susceptibility by

$$\mu = \mu_0 (1 + \chi) \tag{4.1}$$

where μ_0 is the vacuum permeability.

Table 4.1 Types of Iron Oxides

Common name	Formula	Information
Hematite	$\alpha\text{-}Fe_2O_3$	Oldest known. Blood-red. Extremely stable. Important pigment and valuable ore
Magnetite	Fe_3O_4	Ferrimagnetic. Responsible for magnetic properties of rocks. Black. Important iron ore
Maghemite	$\gamma\text{-}Fe_2O_3$	Ferrimagnetic. Red-brown. Occurs in soils. Important magnetic material
	$\beta\text{-}Fe_2O_3$	Rare compound synthesized in the laboratory
	$\varepsilon\text{-}Fe_2O_3$	Rare compound synthesized in the laboratory. Intermediate between hematite and maghemite
Wüstite	FeO	Black. Important intermediate in the reduction of iron ores

When a substance is placed in a magnetic field of strength H, the magnetization M (magnetic dipole moment of the sample per unit volume) is related to H by molar magnetic susceptibility χ (in terms of volume or in terms of mass), as given in

$$M + \chi H \tag{4.2}$$

In a magnetic field, the force that acts on a particle is given by

$$F = \mu_0 \chi V_p H \nabla H \tag{4.3}$$

where V_p is the volume of the particle and ∇H the magnetic field gradient.

Magnetite is the most magnetic of all the naturally occurring minerals on earth. It is a ferrimagnetic material. Ferrimagnetic materials consist of at least two interpenetrating sublattices and antiparallel spins of unequal moments, which results in a net magnetic moment.

Superparamagnetism arises as a result of magnetic anisotropy. Anisotropy arises from the existence of preferred crystallographic directions along which electron spins are most readily aligned and the substance most easily magnetized. Because superparamagnetism depends on the anisotropy constant and the size of the particles, it is usually magnetite (or maghemite) nanoparticles with <20 nm sizes that exhibit this property.

There are several methods to synthesize magnetite in the laboratory scale. Coprecipitation, partial oxidation, and thermal decomposition are to name a few. Among these methods, coprecipitation is the simplest and most efficient chemical method for obtaining magnetic particles (Massart, 1981), however, it often yields a wide range of particle size distribution (Theppaleak et al., 2009).

Magnetite (Fe_3O_4) nanoparticles can be produced via co-precipitation of ferrous and ferric ions in a 1:2 stoichiometric ratio in the presence of a base, usually NaOH, or NH_4OH in an aqueous medium. This technique needs fine adjustment of the pH of the medium, and complete precipitation of Fe_3O_4 is expected in a pH range between 8 and 14 (Laurent et al., 2008). One of the problems about this method is that the Fe^{2+}/Fe^{3+} ratio is subject to change due to oxidation of Fe^{2+} in the presence of oxygen. This requires careful control of experimental conditions where the reaction should be carried out in a completely oxygen-free environment. The overall chemical reaction of Fe_3O_4 formation in an oxygen-free environment may be written as

$$Fe^{2+}(aq) + 2Fe^{3+}(aq) + 8OH^-(aq) \rightarrow Fe_3O_4(s) + 4H_2O(l)$$

Even after the synthesis, magnetite (Fe_3O_4) is not very stable and is sensitive to oxidation, especially at high temperatures. Magnetite can be converted into maghemite (γ-Fe_2O_3) in the presence of oxygen as

$$4Fe_3O_4(s) + O_2(g) \rightarrow 6\gamma\text{-}Fe_2O_3(s)$$

which makes careful control of synthesis and storage conditions very important in synthesis of magnetite.

The main advantage of the coprecipitation reaction is that large amounts of nanoparticles can be synthesized in a short time. However, since the synthesis is largely dependent on the reaction kinetics as driven by supersaturation, the control of particle size distribution is limited.

Magnetite, a ferromagnetic material, exhibits superparamagnetic behavior below a certain size. Superparamagnetic particles do not have a permanent magnetic moment and therefore act like nonmagnetic particles in the absence of an external magnetic field. However, when an external magnetic field is applied, particles exhibit magnetic properties and respond to the field. This on–off mechanism gives these particles unique application possibilities in a variety of areas from material science to medicine. Another advantage of the coprecipitation method is that it yields particles of about 8 nm in diameter, which are well within the superparamagnetic range.

Bare (no surface coating) magnetite nanoparticles precipitate in aqueous medium. In order for these particles to be used in applications, they should be stable in the desired fluid. This is achieved by coating the magnetite surface with an appropriate coating, which can be a surfactant or a polymer suitable to be suspended in aqueous or oil media. If the coprecipitation reaction is carried out in the presence of a suitable surfactant, stable nanofluids may be obtained.

In this experiment, Fe_3O_4 nanoparticles with an average diameter of about 8 nm will be synthesized using coprecipitation method under N_2, at high pH. The synthesized particles will be precipitated to obtain solid magnetic particles or suspended in an aqueous medium with the help of a surfactant (Wooding et al., 1991). Oleic acid will be used as the surfactant. Oleic acid gets attached onto the surface of magnetite from its carboxyl group as shown in Figure 4.3.

When excess surfactant is used in the aqueous medium, a secondary surfactant layer forms around the particles where the hydrophilic head-group of the surfactant points out toward the medium and particles get suspended in the aqueous medium with a bilayer of surfactant around them as shown in Figure 4.4.

After the reaction is carried out, Tiron chelation test will be applied to the obtained solution to determine the magnetite concentration. Tiron chelation test is used to determine the iron concentration, which can then be converted into the real concentration of magnetite. Tiron

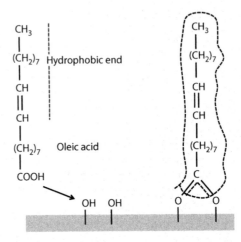

Figure 4.3 The reaction mechanism of oleic acid attachment to iron oxide hydroxyl groups.

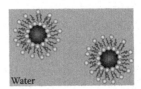

Figure 4.4 Formation of a secondary surfactant layer in the presence of excess surfactants.

(4,5-dihydroxy-1,3-benzenedisulfonic acid disodium salt) chelates with Fe^{3+} and Fe^{2+} ions in a ratio of three Tiron molecules to each iron molecule and exhibits a consistent and strong absorbance at 480 nm for iron solutions with a pH >9.5. In order to liberate the iron content of the magnetite, concentrated hydrochloric acid solution is added to the magnetic fluid. The hydrochloric acid also aids in the removal of the coating around the magnetite. The absorbance of the diluted sample is then measured with a spectrophotometer and converted into concentration of the magnetite using Equation 4.4 (Yuan et al., 2012).

Pre-laboratory questions

1. What is the crystal structure of Fe_3O_4?
2. What is the difference between paramagnetic and superparamagnetic?
3. What are the possible iron oxides that can form upon oxidation of iron?
4. Suggest a method that will differentiate between these iron oxides.
5. Can you suggest applications for magnetic fluids in (a) industrial and (b) medical areas?

Apparatus and chemicals

Iron sulfate heptahydrate ($FeSO_4 \cdot 7H_2O$)
Iron chloride ($FeCl_3$)
Oleic acid ($C_{18}H_{34}O_2$)
Sodium hydroxide (NaOH)
Hydrochloric acid (HCl, 37%)
Tiron ($C_6H_4Na_2O_8S_2$)
Nitrogen (N_2)
Methanol (CH_4O)
Three-necked flask
Mechanical stirrer
A handheld Neodymium magnet

Procedure

Synthesis of iron oxide nanoparticles

1. Place 40 mL of water in a three-necked round-bottom flask.
2. Place a mechanical stirrer through the middle neck and start stirring the liquid.
3. Deaerate the water by passing nitrogen through for about 20 min. You can connect a Pasteur pipette at the end of a tubing which is connected to the N_2 and insert the pipette inside the solution through one of the necks.
4. Weigh 2.41 g of $FeSO_4 \cdot 7H_2O$ (or any Fe^{2+} salt but then the amounts should be calculated accordingly) and 2.82 g of $FeCl_3$ (or any Fe^{3+} salt but then the amounts should be calculated accordingly) separately, add to the deaerated water, and allow the solution to dissolve completely by waiting another 10 min under N_2 flow.
5. Stop the nitrogen flow and quickly add 5 mL of a strong base (5.14 g of NaOH). You will immediately observe blackening of the solution which suggests the formation of the magnetite (Fe_3O_4) phase.
6. Continue stirring the sample for another half an hour.
7. Stop the stirring and pour the solution into a beaker. Add enough methanol to the medium to completely sediment the particles either with a magnet or by centrifugation. (If no strong magnet or centrifuge is available, just allow some time for the particles to sediment.)
8. Wash the particles twice with distilled water.
9. In order to obtain water-soluble particles in water, the same procedure should be repeated using 2 mL of oleic acid as a surfactant to stabilize the particles and the experiment should be carried out at 60°C. The surfactant should be introduced to the medium as suspended in the base. Oleic acid coats the surface of the magnetite and electrostatically stabilized magnetite nanoparticles are obtained.

10. Using a handheld Neodymium magnet, check the response of the particles to see if they are magnetic.
11. The hydrodynamic radius of the water-soluble particles can be determined by DLS (see Chapter 2.5). The core size of bare or coated magnetite can be measured by transmission electron microscopy (TEM; see Chapter 2.6).
12. The crystallographic information on the synthesized particles can be obtained by X-ray diffraction method (see Chapter 2.4).

Tiron test to determine the magnetite concentration in a suspension

1. Prepare 10 mL of 0.083 g/mL Tiron in water solution (Tiron Stock Solution).
2. Take a small amount of magnetic fluid (0.1 mL for a very concentrated fluid) and mix with 0.4 mL of concentrated (37%) hydrochloric acid.
3. Heat the obtained suspension with a heat gun for a few seconds until the color of the solution turns yellow.
4. Allow the mixture to cool down to room temperature and mix with 0.6 mL of Tiron stock solution.
5. Add 3 mL of 4 M sodium hydroxide solution. You will observe a color change to red.
6. Measure the absorbance of this solution using a UV–Vis Spectrophotometer at $\lambda = 480$ nm.
7. If the red solution is too dark to be measured (i.e., absorbance values are >1), dilute the solution by 4 or 8 times using water as the solvent.

Observations and data

Magnetite synthesis	
Color of the solution before the base addition	
Color of the solution after the base addition	
Is there a suspension or a particulate solution at the end of the experiment?	
Do the particles respond to a magnet?	
Tiron test	
Color of the precipitate	
Color of the solution after HCl addition	
Color of the solution after heat gun	
Color of the solution after addition of Tiron	
Color of the solution after NaOH addition	

Calculations

The absorbance (*A*) of the liquid sample is related to the molar extinction coefficient at a given wavelength, ε (L mol^{-1} cm^{-1}), the molar concentration, *c* (mol L^{-1}), and the thickness of the cuvette, *d* (cm), by Beer–Lambert law: $A = \varepsilon c d$. The molar extinction coefficient, ε, is 39.986 L/g mol (Sharpe, 2004). Then the number of moles of $FeCl_3$ in the sample is

$$n_{FeCl_3} = \frac{A}{\varepsilon \cdot d} \cdot \frac{DF \cdot V_{final}}{M_{FeCl_3}}$$

(4.4)

where DF is the dilution factor, and the molecular weight of $FeCl_3$ (M_{FeCl_3}) is 162.2 g/mol.

The stoichiometric coefficient for Fe_3O_4 and $FeCl_3$ is 1:3 from the reaction between iron(III)oxide and hydrochloric acid. The concentration of iron oxide is calculated by multiplying the mole with the molecular weight of Fe_3O_4, followed by division with sample volume (V_{sample}):

$$c_{Fe_3O_4} = \frac{A}{\varepsilon \cdot d} \cdot \frac{DF \cdot V_{final}}{M_{FeCl_3}} \cdot \frac{1}{3} \frac{M_{Fe_3O_4}}{V_{sample}}$$

(4.5)

where the molecular weight of Fe_3O_4 ($M_{Fe_3O_4}$) is 231.53 g/mol.

Clean-up and disposal

1. Collect the solutions in separate waste containers.
2. Clean all the glassware and laboratory station.
3. Wash your hands thoroughly before leaving the laboratory.

Post-laboratory questions

1. For this experiment, the ratio of Fe^{2+} to Fe^{3+} is very important. Why?
2. What is the role of nitrogen in this experiment?
3. What is the role of the surfactant?
4. What part of the surfactant molecule shows an affinity toward the particle surface? Can you suggest another surfactant or polymer that can serve the same purpose?
5. What is the principle behind the determination of iron content using Tiron? Which molecule does the final solution contain?
6. Why is there a need for dilution before the absorbance measurement in the spectrophotometer?

Note for the instructors

It would be time-efficient if some students carry out the synthesis of bare nanoparticles, while others use a surfactant to obtain nanoparticles that are suspended in water. Some of the surfactant-stabilized magnetite particles may settle in solution. The suspended ones can be removed by centrifugation leaving the sedimented ones behind. In this experiment, capric acid and citric acid can also be used as a stabilizer, but because these are in solid form at room temperature, the amounts should be adjusted to 2 g.

References

Laurent, S., D. Forge, M. Port, A. Roch, C. Robic, L. Vander Elst, and R.N. Muller. Magnetic Iron Oxide Nanoparticles: Synthesis, Stabilization, Vectorization, Physicochemical Characterizations, and Biological Applications. *Chemical Reviews* 108(6), 2008: 2064–110.

Massart, R. Preparation of Aqueous Magnetic Liquids in Alkaline and Acidic Media. *IEEE Transactions on Magnetics* 17(2), 1981: 1247–48.

Sharpe, S.A. *Magnetophoretic Cell Clarification*. Massachusetts Institute of Technology, 2004.

Theppaleak, T., G. Tumcharern, U. Wichai, and M. Rutnakornpituk. Synthesis of Water Dispersible Magnetite Nanoparticles in the Presence of Hydrophilic Polymers. *Polymer Bulletin* 63, 2009: 79–90.

Wooding, A., M. Kilner, and D.B. Lambrick. Studies of the Double Surfactant Layer Stabilization of Water-Based Magnetic Fluids. *Journal of Colloid and Interface Science* 144(1), 1991: 236–42.

Yuan, Y., D. Rende, C.L. Altan, S. Bucak, R. Ozisik, and D.A. Borca-Tasciuc. The Effect of Surface Modifications on Magnetization of Iron Oxide Nanoparticle Colloids. *Langmuir* 28(36), 2012: 13051–59.

Experiment 14: Synthesis of silica nanoparticles (hydrolysis condensation)

Purpose

Silica nanoparticles have been widely used in biosensors and biochips. The functionalization of these nanoparticles introduces new properties, such as fluorescence, magnetism, therapeutic ability, and catalytic function, to the material. Silica nanoparticles are usually prepared either by the Stöber method or water-in-oil (w/o) microemulsion method. The aim of this experiment is to synthesize two different sizes of silica nanoparticles with the Stöber method.

Theoretical background

78% of Earth's crust consists of silicon and oxygen compounds, both in amorphous and crystalline form, for example, quartz, flint, opal, silicates, etc. Silicon is not found in the nascent form, but always in combination with oxygen (silica) or hydroxides (silicic acid). Silica can be synthesized by various techniques to prepare spherical nanoparticles, transparent films, or solid flat materials. Silica nanoparticles have received great attention in recent years since they are chemically and thermally stable, have large surface area, and possess great suspension ability in aqueous solutions.

Silica synthesis is governed by the chemical properties of the surface, where silanols and siloxanes are present on the surface. The high silanol concentration on the surface facilitates a wide variety of surface reactions and subsequent binding of biomolecules. The hydroxyl groups on the surface can react with various compounds to form amine, carboxyl, or thiol groups. In addition, silica surface can be decorated with large biomolecules, such as biotin–avidin, antigen–antibodies, proteins, and DNA (Li et al., 2011).

Compared to other nanoparticles, silica nanoparticles possess several advantages, such as easy surface modification. Silica nanoparticles are easy to separate via centrifugation during particle preparation because of their high density (1.96 g/cm^3). Silica nanoparticles are hydrophilic and biocompatible, they are not subject to microbial attack, and no swelling or porosity change occurs with changes in pH (Jin et al., 2009).

Silica nanoparticles are synthesized either by the Stöber method or by the w/o microemulsion method. Stöber's method produces monodispersed spherical particles with sizes covering almost the whole colloidal range, from 50 nm to 2 mm (Stöber et al., 1968; Beganskiene et al., 2004). Silica nanoparticles are prepared by the hydrolysis and polymerization of a precursor, tetraethylorthosilicate (TEOS or $SiO(C_2H_5)_4$), in the ethanolic medium and in the presence of ammonia as the catalyst (Stöber et al., 1968). The hydrolysis of TEOS produces silicic acid, which undergoes a condensation process to form amorphous silica (Green et al., 2003).

The hydrolysis reaction

$$SiO(C_2H_5)_4(l) + H_2O(l) \rightarrow O(C_2H_5)_3Si(OH)(l) + ROH(l)$$

produces an intermediate TEOS monomer ($SiO(C_2H_5)_4$). Following the hydrolysis reaction, the immediate condensation reaction produces silica nanoparticles:

$$(C_2H_5O_3)Si(OH)(l) + H_2O(l) \rightarrow SiO_2(s) + 3C_2H_5OH(l)$$

Like any other synthesis of colloids, the diameter of silica particles from the Stöber process is mainly controlled by the relative contribution from nucleation and growth which are governed by the stoichiometry of ammonia, TEOS, and water.

For the Stöber nanoparticles, surface modification is usually done after nanoparticle synthesis to avoid potential secondary nucleation. Surface modification of microemulsion nanoparticles can be achieved in the same manner or via direct hydrolysis and co-condensation of TEOS and other organosilanes in the microemulsion solution.

Materials

Tetraethylorthosilicate (TEOS)
Ammonia (NH$_3$, 28%)
Tetrahydrofuran (THF)
Ethanol (C$_2$H$_5$OH)
Tridecafluoro-1,1,2,2-tetrahydrooctyl triethoxysilane (F-TEOS)
Vacuum pump with 0.45-μm hydrophobic filtering paper

Procedure

15-nm silica nanoparticles

1. At room temperature, place 194 g of ethanol and 4.99 g of ammonia in a round-bottom flask and mix for at least 15 min.
2. Instantly add 8.62 g of TEOS.
3. Mix for 24 h. As the silica nanoparticles form, the reaction mixture will become white from transparent.
4. Precipitate the nanoparticles with a rotary evaporator at 60°C and 90 rpm for 2 h.
5. To remove any residual solvent, dry the nanoparticles in vacuum at 60°C overnight.
6. Vacuum filter the nanoparticles first with THF and then with distilled water.
7. Dry in vacuum oven at 60°C overnight.

100-nm silica nanoparticles

8. At room temperature, place 87.5 g of ethanol, 4.37 g of deionized water, and 4.12 g of ammonia in a round-bottom flask, and mix for at least 15 min.
9. Instantly add 8.62 g of TEOS.
10. Mix for 24 h. As the silica nanoparticles form, the reaction mixture will become white from transparent.
11. Precipitate the nanoparticles with a rotary evaporator at 60°C and 90 rpm for 2 h.
12. To remove any residual solvent, dry the nanoparticles in a vacuum oven at 60°C overnight.

13. Vacuum filter the nanoparticles first with THF and then with distilled water.
14. Dry in a vacuum oven at 60°C overnight.

Surface modification with silane coupling agent

To make hydrophobic nanoparticles, the silica surface can be modified with silane coupling agents (Goren et al., 2010).

1. Once the reaction is completed after 24 h, add 0.35 g of the silane coupling agent to the reaction mixture and mix the reaction mixture for an additional 24 h.
2. Remove unreacted solvent by a rotary evaporator at 60°C and 90 rpm for 2 h.
3. Dry the nanoparticles in vacuum at 60°C overnight.
4. Vacuum filter the nanoparticles with THF.

After the synthesis

The particle size analysis can be performed via imaging with scanning electron microscopy (SEM) or TEM. The cluster size of the nanoparticles can be confirmed with DLS.

Clean up and disposal

1. Collect the solutions in separate waste containers.
2. Clean all the glassware and laboratory station.
3. Wash your hands thoroughly before leaving the laboratory.

Post-laboratory questions

1. What is the role of ammonia during the synthesis?
2. What is the role of ethanol during the synthesis?
3. How does the surface modifier change the surface hydrophobicity?

Note for the instructor

Ammonia used for the synthesis should be handled very carefully. Upon addition of TEOS, a color change from transparent to white would be observed gradually in 10–15 min. A proper filter paper should be selected during vacuum filtration, and students might observe the success of the experiment (in terms of coating) by the nanoparticle behavior in water as well as during vacuum filtration. The duration of the reaction controls the size of the nanoparticles, and hence the control of the size of 15-nm silica nanoparticles is harder than 100-nm silica nanoparticles. The whole experiment takes a few days to complete although the preparations can be done in a laboratory session. The students should come back to the laboratory to finish off their experiments.

References

Beganskiene, A., V. Sirutkaitis, M. Kurtinaitiene, R. Juskenas, and A. Kareiva. FTIR, TEM and NMR Investigations of Stöber Silica Nanoparticles. *Materials Science* 10, 2004: 287–90.

Goren, K., L. Chen, L. Schadler, and R. Ozisik, Influence of Nanoparticle Surface Chemistry and Size on Supercritical Carbon Dioxide Processed Nanocomposite Foam Morphology. *Journal of Supercritical Fluids* 51, 2010: 420–7.

Green, D.L., J.S. Lin, Y.-F. Lam, M.Z.C. Hu, D.W. Schaefer, and M.T. Harris. Size, Volume Fraction, and Nucleation of Stöber Silica Nanoparticles. *Journal of Colloid and Interface Science* 266(2), 2003: 346–58.

Jin, Y., A. Li, S.G. Hazelton, S. Liang, C.L. John, P.D. Selid, D.T. Pierce, and J. Xiaojun Zhao. Amorphous Silica Nanohybrids: Synthesis, Properties and Applications. *Coordination Chemistry Reviews* 253(23–24), 2009: 2998–3014.

Li, S., J. Singh, H. Li, and I.A. Banerjee. *Biosensor Nanomaterials*. Weinheim: Wiley-VCH, 2011.

Stöber, W., A. Fink, and E. Bohn. Controlled Growth of Monodispersed Silica Spheres in the Micron Size Range. *Journal of Colloid and Interface Science* 26(1), 1968: 62–69.

Experiment 15: Synthesis of nanoparticles in microemulsions (reduction in a confined environment)

Purpose

The purpose of this experiment is to prepare a w/o microemulsion using an anionic surfactant and then use this microemulsion as a nanoreactor to synthesize nickel (Ni) nanoparticles.

Theoretical background

Microemulsions

Microemulsions are thermodynamically stable isotropic dispersions of oil, water, and surfactant. Emulsions with large droplet diameters are inherently unstable due to high interfacial tension between oil and water. Surfactants lower this interfacial tension and somewhat stabilize the emulsions. However, because emulsions are not thermodynamically stable, they tend to phase separate over time. By carefully choosing the surfactant and adjusting the ratio of the oil, water, and the surfactant, the droplet size may spontaneously (without external energy input) decrease below a certain diameter and a thermodynamically stable microemulsion may form. Often, a cosolvent such as a medium carbon chainlength alcohol is required to obtain a microemulsion. However, ternary systems of oil, water, and the surfactant have also been shown to form microemulsions and have been more widely used in applications due to their simplicity.

The droplet size in microemulsions is in the range 5–50 nm. Owing to the small size of the dispersed phase, the dispersion appears transparent and resembles a solution.

As opposed to being thought of as small emulsions, microemulsions can also be considered as swollen micelles. Whether oil-in-water (o/w) or w/o, microemulsions can be visualized as enlarged micelles (or reversed micelles).

Microemulsion droplets diffuse and collide with each other, often resulting in a short-lived dimer formation followed by an exchange of the droplet contents. This exchange takes place rather fast, in the order of milliseconds (Fletcher et al., 1987; Clark et al., 1990).

Microemulsions are traditionally classified based on the type of phase equilibria (Figure 4.5) (Winsor, 1948).

Winsor I: This is the o/w microemulsion where the surfactant is water-soluble. In this system, the surfactant is present in the oil phase at very small concentrations and excess oil phase is in coexistence with the surfactant-rich o/w microemulsion phase.

Winsor II: This is the w/o microemulsion where the surfactant is oil-soluble. In this system, the surfactant is present in the aqueous phase at very small concentrations and the excess aqueous phase is in coexistence with the surfactant-rich w/o microemulsion phase.

Winsor III: This is a three-phase system where a surfactant-rich middle phase (microemulsion) coexists with both aqueous and oil phases. The middle phase is also called the bicontinuous phase.

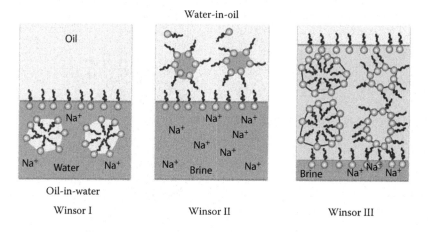

Figure 4.5 Winsor-type system.

Winsor IV: These systems can be either o/w or w/o microemulsions where an isotropic micellar solution forms upon addition of usually a cosurfactant such as alcohol.

Depending on the surfactant type and sample environment, Winsor I, II, III, or IV may form. Phase transitions between these systems may take place upon increasing electrolyte concentration (in the case of ionic surfactants) or temperature (for nonionics).

When preparing a microemulsion, a semiempirical formula is used to determine the droplet size of the microemulsion:

$$r_{hydrodynamic} = r_{core} + l_{tail} \tag{4.6}$$

where $r_{hydrodynamic}$ is the hydrodynamic radius, r_{core} the droplet size, and l_{tail} the length of the hydrophobic tail of the surfactant. The droplet size is related to the dimensions of the surfactant and the w_0 value, the water-to-surfactant molar ratio ([H$_2$O]/[surfactant]). Assuming spherical head-group and that all the surfactant molecules go to the water–oil interface, the relationship between r_{core} and w_0 is given as

$$r_{core} = \frac{3V_m}{a_0} w_0 \tag{4.7}$$

where a_0 is the headgroup area of the surfactant, V_m the volume of the hydrophobic region of the surfactant, and V_m/a_0 a constant unique to each surfactant.

In order to prepare a w/o microemulsion of a given droplet size, dioctyl sodium sulfosuccinate (commonly known as AOT) will be used as a surfactant and heptane will be used as the oil phase. The chemical structure of AOT is shown in Figure 4.6.

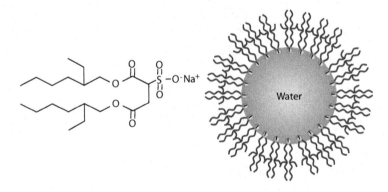

Figure 4.6 Structure of dioctyl sodium sulfosuccinate.

For AOT, the hydrodynamic radius is given as (Nicholson and Clarke, 1984)

$$r_{hydrodynamic} = 0.175w_0 + 1.5 \tag{4.8}$$

For example, if one wants to prepare a 50-mL microemulsion with a droplet core radius of 3.25 nm using 0.1 M AOT as the stabilizer, the value of w_0 should be taken as 10. For [AOT] = 0.1 M, 0.05 mol H_2O (0.9 mL of H_2O) is calculated to be present in 50 mL of solution.

In this experiment, in the water compartment of the microemulsion, Ni^{2+} cations will be reduced by sodium borohydrate, $NaBH_4$, which is a highly reactive base and reducing agent, to obtain Ni nanoparticles employing two different routes (Chen and Wu, 2000; Chen and Hsieh, 2002).

Synthesis of nanoparticles in microemulsion droplets

Because microemulsions provide a confined space for the aqueous reactions to occur, they are ideal for production of nanoparticles. When two reactants of a reduction reaction are dissolved in two separate but identical w/o microemulsions and mixed afterwards, they will form a precipitate as the product. Then, the surfactant can be washed off from the surface of the precipitated particles. This is shown in Figure 4.7.

The reaction will be carried out as follows:

$$2Ni^{2+}(aq) + N_2H_4(aq) + 4OH^-(aq) \rightarrow 2Ni(s) + N_2(g) + 4H_2O(l)$$

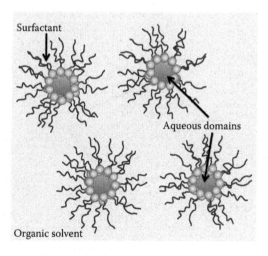

Figure 4.7 Making particles in water droplets.

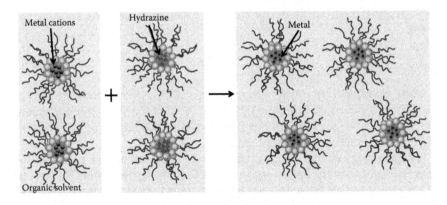

Figure 4.8 Reaction by droplet collision.

There are two routes this reaction can be carried out.

Route 1: Placing each reactant (metal ion and sodium borohydride (NaBH$_4$)) in two different microemulsions and allowing the reaction to take place as a result of droplet collision (Figure 4.8).

Route 2: Placing the metal ion in the microemulsion and adding sodium borohydride (NaBH$_4$). Sodium borohydride aqueous solution will quickly diffuse into the water compartment where the reaction will take place (Figure 4.9).

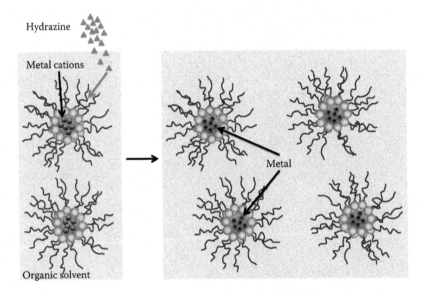

Figure 4.9 Reaction by external sodium borohydride (NaBH$_4$) addition.

After carrying out the metal reduction reaction in microemulsion droplets following the two routes, the size of the obtained particles and the rate of nanoparticle formation will be compared.

Pre-laboratory questions

1. What is the type of microemulsion that is likely to form when the surfactant is water-soluble?
2. Why do microemulsions appear transparent?
3. Does the droplet size of a microemulsion change upon dilution? Why?

Apparatus and chemicals

Volumetric flasks
Magnetic stirrer
Dioctyl sodium sulfosuccinate (AOT)
n-Heptane (C_7H_{16})
Water
Nickel (II) sulfate ($NiSO_4$)
Sodium borohydride ($NaBH_4$)

Procedure

1. Prepare a 100-mL stock solution of 0.5 M AOT in n-heptane. Because AOT is very sticky, a particular procedure for preparation is suggested: Place the required amount of AOT in a 250-mL beaker with about 50 mL of heptane inside. Cover the beaker with an aluminum foil and sonicate the solution to dissolve AOT. Some heating due to sonication is acceptable. After all the AOT is dissolved, transfer the solution into a volumetric flask. Rinse the beaker with small aliquots of heptane and pour them into the volumetric flask to reach 100 mL.
2. Prepare 10 mL of 0.5 M stock Ni solution in water (all students can use this stock solution).
3. Each group can prepare a different droplet size ($w_o = 5$, 10, 15, 20). Calculate the amount of water phase you will need for preparing 50 mL of a microemulsion solution containing 0.1 M AOT.
4. Place the calculated amount of aqueous Ni solution (water phase) and add a little n-heptane while stirring magnetically.
5. Keep on adding n-heptane until the solution becomes clear. Then fill up to the 50-mL mark with n-heptane.
6. Separate this solution into two equal parts (label them as A1 and A2).
7. Prepare 10 mL of 0.1 M $NaBH_4$ aqueous stock solution.
8. Prepare another 50 mL of microemulsion, using 0.02 M sodium borohydride ($NaBH_4$) ($V_{sodium\ borohydride} = V_{aqueous\ phase}$) as the water phase using the same procedure. Label this as B.

9. Mix 5 mL of A1 with 5 mL of B and record the time needed for Ni nanoparticle formation. Repeat this at least three times.
10. Take 5 mL of A2. This time add one-fifth of the calculated amount of $NaBH_4$ (0.1 M) dropwise (slowly) and record for Ni nanoparticle formation. Repeat this at least three times.
11. Each time you will observe the Ni nanoparticle formation and eventual precipitation of particles. Add some ethanol to both dispersions to destabilize the microemulsion and precipitate all the nanoparticles.
12. Measure the size of A1 microemulsion using DLS.
13. Measure the size of Ni nanoparticles obtained by two different routes using TEM.

Observations and data

Routes	Color change	Time
1		
2		

Clean up and disposal

1. Collect the solutions in separate waste containers.
2. Clean all the glassware and laboratory station.
3. Wash your hands thoroughly before leaving the laboratory.

Post-laboratory questions

1. Is there a relationship between the size of the droplets and the size of the obtained Ni nanoparticles? Why?
2. What is the role of sodium borohydride ($NaBH_4$) in the reaction? Can you suggest an alternative chemical?
3. What is the reason behind the difference in the rate of nanoparticle formation between the two routes?

Note for the instructor

Preparation of AOT stock solution in heptane takes about 1–2 h. We recommend this solution to be prepared before the laboratory session.

Microemulsions can be formed with other ternary systems also. Any microemulsion can be used in this experiment. However, if a different surfactant is used, the droplet size should be calculated using the values of a_0, V_m, and l_{tail} for that particular surfactant. Silver metal can also be reduced in microemulsion following a similar procedure. If Experiment 13 was performed by the students before, the reaction rates of reduction in the bulk and in the microemulsion can be compared.

References

Chen, D.-H. and C.-H. Hsieh. Synthesis of Nickel Nanoparticles in Aqueous Cationic Surfactant Solutions. *Journal of Materials Chemistry* 12(8), 2002: 2412–15.

Chen, D.-H. and S.-H. Wu. Synthesis of Nickel Nanoparticles in Water-in-Oil Microemulsions. *Chemistry of Materials* 12(5), 2000: 1354–60.

Clark, S., P.D.I. Fletcher, and X. Ye. Interdroplet Exchange Rates of Water-in-Oil and Oil-in-Water Microemulsion Droplets Stabilized by Pentaoxyethylene Monododecyl Ether. *Langmuir* 6(7), 1990: 1301–9.

Fletcher, P.D.I., A.M. Howe, and B.H. Robinson. The Kinetics of Solubilisate Exchange between Water Droplets of a Water-in-Oil Microemulsion. *Journal of the Chemical Society, Faraday Transactions 1: Physical Chemistry in Condensed Phases* 83(4), 1987: 985–1006.

Nicholson, J.D. and J.H.R. Clarke. In *Surfactants in Solution*, edited by K.L. Mittal and B Lindman. 1671. New York: Plenum Press, 1984.

Winsor, P.A. Hydrotropy, Solubilisation and Related Emulsification Processes. *Transactions of the Faraday Society* 44, 1948: 376–98.

Experiment 16: Synthesis of latex particles (in situ polymerization in miniemulsions)

Purpose

The purpose of this experiment is to use an o/w miniemulsion as a nanoreactor to carry out a polymerization reaction and obtain latex nanoparticles.

Theoretical background

Polymeric dispersions are widely used in many applications ranging from paints to drug-delivery systems. Besides their unique properties that make them desirable for a variety of applications, the need for waterborne products as opposed to solvent-based systems due to environmental concerns lead to the development of new pathways to obtain polymeric dispersions.

The most conventional method of obtaining polymeric dispersions (latex) is emulsion polymerization. Although this term gives the impression that the polymerization takes places in the emulsion droplet, in reality polymerization takes place in the micelles that are present in the medium besides the emulsion droplets. In emulsion polymerization, the monomer forms the oil phase of the emulsion. Surfactants stabilize the monomer droplet. Above the critical micelle concentration of the surfactant, there are also micelles present in the medium as shown in Figure 4.10. These micelles get swollen by monomers. Nucleation may take place in three different ways (Schork et al., 2005):

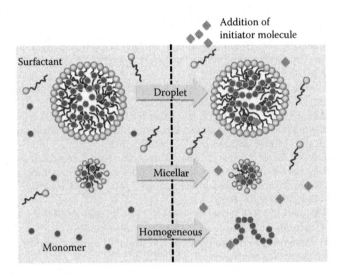

Figure 4.10 Schematic representation of emulsion polymerization in a droplet, micelle, and aqueous medium in the presence of an initiator.

1. The free radicals in the aqueous phase initiate polymerization inside the micelles (micellar nucleation).
2. The free radicals in the aqueous phase initiate polymerization inside the monomer droplet (droplet nucleation).
3. Aqueous-phase radicals polymerize to form oligomers. These continue to grow until they reach a critical chain length, the size of a primary particle, and then precipitate. Throughout the growth process, the oligomers may also flocculate or coagulate (homogeneous nucleation).

These mechanisms compete and coexist in the same system and often one dominates. Owing to the large size of monomer droplets with respect to micelles and their consequent low interfacial area, this mechanism often cannot compete for water-borne free radicals. And when it happens, they lead to the formation of a few very large particles which can be neglected. Homogeneous nucleation predominates when the water solubility of the monomer is high.

Emulsion polymerization can be summarized as shown in Figure 4.10.

Micellar nucleation predominates when the aqueous solubility of the monomer is low. Once the polymerization starts, the core of the micelles is depleted of monomers and they are supplied to the micelles from the monomer droplets. These insoluble monomers should pass through the aqueous medium to reach the micelle core, which presents a problem.

This problem can be remedied if the micellar nucleation can be suppressed and droplet nucleation can be favored. This is possible if the monomer droplets and the micelles prepared are of comparable size, creating a large interfacial area of the droplets. This creates the need for miniemulsions where the droplet sizes are below 100 nm and are stable for sufficient amount of time to allow polymerization. The surfactants in the medium are used to support the large interfacial area, leading to a shift in the droplet–micelle equilibrium. In a properly formulated miniemulsion, almost all the surfactants can be considered to be around the droplets. Therefore, not only do the small droplets compete effectively with micelles, but their presence also causes the dissolution of the micelles, leaving droplet nucleation as the dominant particle nucleation process.

A typical miniemulsion formulation includes water, a monomer (or a mixture of monomers), a costabilizer, a surfactant, and an initiator. Miniemulsions can be prepared by applying shear, using a homogenizer or an ultrasonicator (Asua, 2002).

In this experiment, an o/w miniemulsion dispersion will be used to obtain latex particles via miniemulsion polymerization. The miniemulsion will be prepared using an ultrasonicator or homogenizer, and a water-soluble initiator will initiate the polymerization reaction inside the miniemulsion droplet. At the end of the experiment, latex particles will be collected and their sizes will be determined.

Pre-laboratory questions

1. What are the three possible nucleation sites in emulsion polymerization?
2. Under which conditions does the droplet nucleation become dominant?
3. What are the roles of the costabilizer, the surfactant, and the initiator in emulsion polymerization?

Apparatus and chemicals

Styrene (C_8H_8)
Sodium dodecyl sulfate (SDS)
Hexadecane ($C_{16}H_{34}$)
Potassium persulfate ($K_2S_2O_8$)
Nitrogen (N_2)
Magnetic stirrer
Four-necked round-bottom flask equipped with a paddle stirrer, a thermometer, a nitrogen inlet, and a reflux condenser (reactor)
Hallow fiber membrane
Peristaltic pump

Procedure

1. Mix 160 mL of H_2O, 32 mmol hexadecane, 3.2 mmol SDS, and 320 mmol styrene using a magnetic stirrer for 10 min.
2. Homogenize the emulsion for 15 min under vigorous stirring by using an ultrasonicator or homogenizer. Use an ice bath during sonication to avoid heating.
3. Transfer the solution to the reactor. (Take out 5 mL of the sample and save for later measurements.)
4. Purge the system with nitrogen for 15 min, while stirring at 100 rpm to remove oxygen.
5. Inject 5 mL of 1.6 mmol solution of potassium persulfate.
6. To initiate the reaction, heat the solution to 60°C at a 200 rpm stirring rate and continue the reaction for 6 h.
7. After 6 h, filter and wash the polymer latex nanoparticles using a Buhner funnel with a filter and deionized water.
8. Dry the obtained nanoparticles in vacuum at 60°C overnight.
9. The miniemulsion (5 mL of the sample) droplet size can be determined using DLS.
10. The obtained particles can be visualized using SEM or their size distribution can be determined using DLS.

Observations and data

Color of the emulsion	
Transparency of the emulsion	
Color of the precipitated particles	

Post-laboratory questions

1. Why is the reaction carried out under nitrogen?
2. What is the role of hexadecane?
3. What is the role of potassium persulfate?
4. What happens when the solution is heated to 60°C after the addition of the initiator?
5. How does the size of the miniemulsion droplet and the obtained latex particles compare?

Note for the instructor

The synthesis of latex nanoparticles cannot be completed in a single laboratory session. Students will have to come back to finish their experiments.

References

Asua, J.M. Miniemulsion Polymerization. *Progress in Polymer Science* 27(7), 2002: 1283–346.

Schork, F.J., Y. Luo, W. Smulders, J.P. Russum, A. Butté, and K. Fontenot. Miniemulsion Polymerization. Chap. 2. In *Polymer Particles.* Edited by M. Okubo. *Advances in Polymer Science,* 129–255. Berlin: Springer, 2005.

chapter five

Applications

Experiment 17: Food colloids

Purpose

Food colloids, such as ice cream, salad dressings, condiments, and mayonnaise, represent a significant portion of our diet. As an everyday example, mayonnaise, which is an oil-in-water (o/w) emulsion, can be prepared in the presence or absence of an emulsifier.

Theoretical background

Food emulsions are broadly categorized as: (i) o/w emulsions, where droplets of oil are suspended in an aqueous continuous phase and (ii) water-in-oil (w/o) emulsions. The former group exists in many forms (creamers, ice cream, mayonnaise) and their properties can be controlled by surfactants and components present in the aqueous phase. On the other hand, butter, margarines, and fat-based spreads are examples of w/o emulsions (Le Révérend et al., 2010).

Food emulsions are also categorized depending on their end uses and stability: some are already end products, such as coffee creamers, and their shelf life is comparatively longer. On the other hand, some emulsions are also used as ingredients, which participate in other food preparations. The control and processing of the latter is complex because these emulsions should interact with other food products without compromising their stability (Dalgleish, 2006).

Conventional methods of preparing emulsions use several types of mixing equipment, homogenizers, ultrasonics, or microfluidizers. Majority of the food products have flavors, which are compounded in the dispersed phase. However, strong shear stress in these conventional preparation methods lead to coalescence of dispersed phase and formation of polydispersed emulsions, which limits to control the droplet size (Sotoyama et al., 1999). In addition, the dispersed phase size is important to prevent bacterial growth. In small droplet sizes, since there are limited nutrients available, the bacterial growth is minimized. Recent advances in the food emulsions offer a new technique, called membrane emulsification, in which low pressure is applied to disperse phase to direct it to continuous phase through a membrane. The primary feature of this technique is the control of dispersed-phase droplet sizes by selecting an appropriate membrane pore size (Charcosset, 2009).

Mayonnaise is essentially an o/w emulsion in which the dispersed (oil) phase represents approximately 80% of the emulsion.

Egg (a medium-sized one) contains 9% fat and 12.6% protein. Most of the fat is in the egg yolk. Egg yolk also contains lecithin, which is a phospholipid molecule. Each lecithin molecule contains a polar end that is attracted to water and a nonpolar end that is attracted to oil. The result is that the lecithin dissolves half of itself in water and the other half in oil. Hence, egg yolk serves also as an emulsifier and stabilizes the emulsion.

The aim of this experiment is to prepare an o/w emulsion (mayonnaise) in the absence or presence of an emulsifier (egg yolk). The water phase of this emulsion can be vinegar or lemon juice.

Materials

1 egg yolk
Vinegar (or lemon juice)
Vegetable oil
Mechanical stirrer or homogenizer

Procedure

Preparation of mayonnaise without emulsifier
1. Add 8 mL of vinegar in the homogenizer or ultrasonicator.
2. Add small portions of oil (total of 250 mL) and homogenize the emulsion under vigorous stirring until an emulsion is formed. Or instead of using a homogenizer, a vigorous mechanical stirrer would also work.
3. Add another 8 mL of the vinegar and continue to homogenize the solution.
4. Set the mayonnaise aside and record your observations over time.

Preparation of mayonnaise with emulsifier
1. Add 8 mL of vinegar and an egg yolk in the homogenizer or ultrasonicator.
2. Mix the egg yolk and vinegar until it thickens.
3. Add the oil in small portions, simultaneously while mixing.
4. Add another 8 mL of the vinegar and continue to homogenize the solution.
5. Set the mayonnaise aside and record your observations over time.

Observations

	$t = 0$			$t = 30$ min		
	Color	Texture	Smell	Color	Texture	Smell
Without emulsifier						
With emulsifier						

Clean-up and disposal

1. In this experiment, raw eggs are used; they may contain *Salmonella*, a bacterium that can cause food-borne illness. Do not consume the food prepared.
2. Discard the emulsions.
3. Clean all the glassware and laboratory station.
4. Wash your hands thoroughly before leaving the laboratory.

Post-laboratory questions

1. What is the role of egg yolk?
2. Find the structure of lecithin and identify the water- and oil-soluble parts.
3. What other ingredients are being used as emulsifiers?

Note for the instructor

It is important to add the oil of about 1–2 mL at a time at the beginning. Then, it is possible to add the oil more quickly while homogenizing to incorporate the oil into the emulsion.

References

Charcosset, C. Preparation of emulsions and particles by membrane emulsification for the food processing industry. *Journal of Food Engineering* 92(3), 2009: 241–49.

Dalgleish, D.G. Food emulsions—their structures and structure-forming properties. *Food Hydrocolloids* 20(4), 2006: 415–22.

Le Révérend, B.J.D., I.T. Norton, P.W. Cox, and F. Spyropoulos. Colloidal aspects of eating. *Current Opinion in Colloid & Interface Science* 15(1–2), 2010: 84–9.

Sotoyama, K., Y. Asano, K. Ihara, K. Takahashi, and K. Doi. Water/oil emulsions prepared by the membrane emulsification method and their stability. *Journal of Food Science* 64(2), 1999: 211–15.

Experiment 18: Body wash formulation

Purpose

To demonstrate the use of surfactants to prepare a body wash and learn the function of each ingredient in the formulation.

Theoretical background

One of the main areas of applications of surfactants is the field of detergents. This term is used for products that promote the removal of material from a surface (e.g., soil from fabric or food from a dish) and disperse and stabilize materials in a bulk matrix (e.g., suspension of oil droplets in a mobile phase like water). Depending on where the detergent will be used, what kind of performance is expected, and economics and environmental considerations, an appropriate formulation is developed.

The most common detergents are those used in household cleaning and personal care. These products are:

1. Laundry detergents and laundry aids
2. Dishwashing products
3. Household cleaning products
4. Personal cleansing products

Modern detergents can be composed of 20 or more ingredients depending on the expectations of the final product. The more basic components of detergents are listed below (Showell, 2005).

Surfactants

Surfactants compose the most common ingredient of detergent formulations. The purpose of the presence of surfactants is to promote the dispersion of one phase in the other by modifying the interface of the different phases present. Surfactants reduce the surface tension, hence increase wetting and also create compartments to solubilize oil (dirt) in an aqueous medium through micellization.

Surfactants are generally classified according to their headgroup: Nonionic, anionic, cationic, and amphoteric (both anionic and cationic characteristics). The surfactants used in detergent formulations are generally expected to perform in hard water (in the presence of Ca^{2+} and Mg^{2+} ions) and cool wash temperatures. Although most surfactants posses a single alkyl chain, it has been observed that branching of the alkyl chain promotes increased solubility in cold and hard water.

For systems where water is not the bulk phase, specialty surfactants should be used.

Dispersing polymers

Although surfactants act both as solubilizing agents for dirt and as dispersants in the bulk phase, polymeric dispersants are also added in detergent formulations to help in particle suspension. These polymers can possess an ionic group where stabilization of the particles is achieved by electrostatic repulsion or they may possess a nonionic group where the particles are stabilized by steric hindrance.

Builders and chelants

Builder is a general term used for all materials that function as a Ca^{2+} and Mg^{2+} scavenger in an aqueous solution. The presence of these ions (hard water) may cause the precipitation of anionic surfactants, which are widely used in detergent formulations. Chelants are used as builders that bind the metal ions and help remove them from the medium.

Bleaching systems

Bleach is generally used in laundry, automatic dishwashing, and hard surface cleaning detergent formulations. Their path of action is through destroying chromophoric groups responsible for color in "dirt" via oxidative attack.

Solvents

The most common and cheapest solvent for detergent formulations is water. Water is also the most environmentally friendly solvent. On the other hand, some additives incorporated in the detergent formulations have poor solubility in water and therefore require addition of a co-solvent and/or hydrotrope (an organic compound that increases the ability of water to dissolve other molecules).

If for certain applications (such as dry cleaning), water must be avoided, then volatile solvents are used and the ingredients are chosen accordingly.

Performance-enhancing minor ingredients

Minor ingredients can be summarized as follows (Rowe et al., 2009):

1. *Enzymes*—promote dirt removal by the catalytic breakdown of some components.
2. *Whiteners*—used to enhance the visual appearance of white surfaces or fabrics.

3. *Foam boosters*—consumers usually expect large amounts of stable foam (usually in shampoos and dishwashing detergents); specific boosters may be added to the formulations.
4. *Antifoam agents*—some applications like automatic dishwashing foam may interfere with the performance of the machine, and therefore antifoam agents may be required.
5. *Thickeners*—Depending on the application, again consumer needs to determine the expected rheology of the product, so thickeners are used to achieve more viscous products.
6. *Soil release polymers*—these are polymers that alter surface polarity and therefore decrease adherence of soil to a surface, enhancing the removal of dirt.

In this experiment, a body wash will be prepared using a variety of components and their function will be explored (Duman, 2006).

Materials

Trade Name	Content (wt%)	International Nomenclature of Cosmetic Ingredients (INCI) Name
TEXAPON N 70	12.0	Sodium laureth sulfate
DEHYTON PK 45	5.00	Cocamidopropyl betaine
TEXAPON SB 3 KC	3.00	Disodium laureth sulfosuccinate
EUROGLYC D	3.00	Disodium cocoamphodiacetate
GLYCERINE	4.00	Glycerine
COMPERLAN COD	1.50	Cocamide DEA
ISOCIDE B1	0.10	Methylchloroisothiazolinone, methylisothiazolinone, benzyl alcohol
POLYQUART 701 NA	0.60	Polyquaternium-7
D-PANTHENOL	0.30	D-Panthenol
PERFUME	0.70	Perfume
EDTA	0.10	Ethylenediaminetetraacetic acid
D.I. WATER	69.7	Water
Total	100.0	

Materials and procedure

1. Weigh the required amount of TEXAPON N-70 into a 250-mL beaker and add 10 mL of water, stirring magnetically.
2. Weigh the required amounts of DEHYTON PK 45, TEXAPON SB 3 KC, and EUROGYLG D into the beaker one after the other under stirring while adding 25 mL of water.

3. Add the required amounts of GLYCERINE and DEHYTON PK 45 under stirring.
4. Add the required amount of POLYQUART 701 NA and mix the product for 2–3 min.
5. Dissolve the required amount of EDTA in 5.0 mL of water, and add to the mixture.
6. Add the required amount of ISOCIDE B1 and mix until homogeneous.
7. Add the required amounts of D-PANTHENOL and PERFUME and mix until homogeneous.
8. Check the pH.
9. Add citric acid solution (30%) to adjust the pH to average 5.5 (limit 5.0–6.0). Mix and check the pH of the product.
10. Record the amount of citric acid solution added.
11. Add the remaining amount of water until the total amount of the product reached 100 g.
12. Check the pH again and adjust the pH to 5.5 if necessary.

Purpose of ingredients

1. To explore the effect of POLYQUART 701 NA, prepare the same composition without POLYQUART 701 NA (replace the amount with water) and observe the consistency.
2. To explore the effect of EUROGLYC D, prepare the same composition without EUROGLYC D (replace the amount with water) and observe the foam.
3. To explore the effect of GLYCERINE, prepare the same composition without GLYCERINE (replace the amount with water) and compare the feeling on your skin.

Post-laboratory question

1. Write down in the table below the purpose of the presence of each component in the formulation.

Trade Name	International Nomenclature of Cosmetic Ingredients (INCI) Name	Function
TEXAPON N 70	Sodium laureth sulfate	
DEHYTON PK 45	Cocamidopropyl betaine	
TEXAPON SB 3 KC	Disodium laureth sulfosuccinate	
EUROGLYC D	Disodium cocoamphodiacetate	
GLYCERINE	Glycerine	
COMPERLAN COD	Cocamide DEA	
ISOCIDE B1	Methylchloroisothiazolinone, methylisothiazolinone, benzyl alcohol	

continued

Trade Name	International Nomenclature of Cosmetic Ingredients (INCI) Name	Function
POLYQUART 701 NA	Polyquaternium-7	
D-PANTHENOL	D-Panthenol	
PERFUME	Perfume	
EDTA	Ethylenediaminetetraacetic acid	
D.I. WATER	Water	

2. How does the mixture change in terms of consistency, foaming, etc. after the addition of each ingredient?

Note for the instructor

In cosmetic formulations, ingredients that are listed may have different trade names. There may be a different trade name for a similar chemical compound. The wt% contents are given as the wt% of the product mentioned by the trade name, which does not necessarily match the wt% of the pure chemical product.

References

Duman, G. *Pharmaceutical Technology Lab. 1: Dosage Forms and Contemporary Practice.* Faculty of Pharmacy, Yeditepe University, 2006. In this whole chapter, it is cited. Adapted with permission.

Rowe, R.C., P.J. Sheskey, W.G. Cook, and M.E. Fenton. *Handbook of Pharmaceutical Excipients.* 6th Edition. Pharmaceutical Press, London, UK, 2009.

Showell, M. *Handbook of Detergents, Part D: Formulation (Surfactant Science).* Boca Raton, FL: CRC Press, 2005.

Experiment 19: Cream formulations

Purpose

Emulsions are frequently used in medicine to deliver drugs or molecules. When they are directly applied on the skin, they are called as topical medications, for example, creams, foams, gels, lotions, and ointments. Ointments are homogeneous and viscous, applied directly on the skin and mucous membrane. The purpose of this experiment is to understand the fundamental differences between ointments and creams and prepare two types of ointments: o/w and water-soluble.

General introduction to ointments

Ointments are semisolid nonaqueous preparations, whereas creams are ointments with water (Monzer, 2010). They are applied topically onto the skin. Ointments and creams are used for several purposes, such as protectants, antiseptics, emollients, antipruritics, kerotolytics, and astringents. These purposes define the choice of the ointment base. For instance, the desired action, the characteristics of the medication incorporated with the base, stability, and shelf life of the end product.

Ointments have emollient and occlusive characteristics, where the former shows its ability to enhance the hydration of the skin by reducing evaporation and the latter indicates the ointment's ability to block the surface to air penetration.

There are four classes or types of ointment bases, which are classified on the basis of their physical composition (Thompson, 2009). These are:

Hydrocarbon bases
Absorption bases
Water-removable bases
Water-soluble or water-miscible bases

Hydrocarbon-based ointments are advantageous due to their inexpensive processing, nonreactivity, protectant, and water proof, which allows the ointment to remain on the skin for prolonged times and keeps the medication in contact with the skin. However, when desired they cannot be removed easily by water and they have limited capacity of water and alcohol absorption.

Absorption bases include mainly two subgroups: anhydrous absorption bases and w/o emulsions. Anhydrous absorption bases are primarily hydrocarbon bases but contain emulsifiers to stabilize the structure. These bases also exhibit protective, occlusive, and emollient properties. Like hydrocarbon-based ointments, they cannot be easily removed from the skin, and hence holds the medication in contact with the skin. Unlike hydrocarbon-based ointments, they have the capacity to absorb significant amounts of water. However, their greasy and sometimes sticky appearance is not appealing. Some medications in the ointments might have compatibility issues with the emulsifiers added. In general, water-containing ointments might exhibit stability problems as well as provide a suitable environment for bacterial growth.

Water-removable bases are o/w emulsions, and hence can be cleaned with water. One advantage of this type of ointments is their ability to absorb water, and they also allow fluid dissipation. However, they are less protective, less emollient, and less occlusive compared to hydrocarbon and absorption bases. Since they contain water, a preservative should be added to prevent bacterial growth. The continuous phase in o/w emulsions is water, and hence this kind of ointment is prone to drying due to evaporation.

Water-soluble bases are greaseless ointments. Polyethylene glycol is the most used base in the class. This kind of ointment can be easily removed from the skin, without leaving any oil residue. They can absorb some amount of water; however, too much water may dissolve the ointment. They exhibit almost no emollient property, and when incorporated with medications, they might have compatibility problems. Like water-removable ointments, a preservative is required to prevent bacterial growth.

Each ointment base type has different physical characteristics and therapeutic uses based on the nature of its components. Table 5.1 summarizes the composition, properties, and common uses of each of these types (Duman, 2006; Thompson, 2009; Shrewbury, 2012).

Pre-laboratory questions

1. Why does a certain type of cream may be required over the other? Give examples for each ointment type.
2. How are the ointments classified according to physical composition?

Table 5.1 Properties of Ointment Bases

Base Type	Hydrocarbon	Anhydrous	Absorption-Based Water-in-oil emulsions	Oil-in-water emulsions	Water-soluble
Water content	Anhydrous	Anhydrous	Hydrous	Hydrous	Hydrous
Affinity for water	Hydrophilic	Hydrophilic	Hydrophilic	Hydrophilic	Hydrophilic
Water absorption	Cannot absorb water	Can absorb water	Can absorb water (limited)	Can absorb water	Can absorb water (limited)
Solubility in water	Insoluble	Insoluble	Insoluble	Insoluble	Soluble
Wash ability	Not washable	Not washable	Poorly washable	Washable	Washable
Drug release	Poor	Poor	Fair–good	Fair–good	Good
Uses	Protectant	Protectant	–	–	–
Uses	Emollient	Emollient	Emollient	–	–
Uses	Occlusive	Occlusive	Occlusive	Nonocclusive	Nonocclusive
Examples	White petrolatum, vaseline	Lanolin, aquaphor, aquabase	Hydrous lanolin, cold cream	Vanishing cream, dermabase	Polyethylene glycol, polybase

Materials

Sodium dodecyl sulfate (SDS)
Propylene glycol ($C_3H_8O_2$)
Stearyl alcohol ($C_{18}H_{38}O$)
White petrolatum
Polyethylene glycol 400 ($C_{2n}H_{4n+2}O_{n+1}$)
Polyethylene glycol 3350 (Carbowax 3350)
Water

Procedure

Cream type I: Oil-in-water emulsion

1. Calculate the amount of ingredients you will need to prepare 10 g of o/w ointment of composition 1 wt% SDS, 12 wt% propylene glycol, 25 w% stearyl alcohol, 25 wt% white petrolatum, and 37 wt% water.
2. Melt stearyl alcohol and white petrolatum (oil phase) in a hot water bath (around 70°C). If you would like to add a fragrant to the cream (essential oil, e.g., lavender), you can mix the same with the oil phase.
3. Dissolve the remaining ingredients in water and heat the solution to 70°C, again in the hot water bath.
4. Add the oil phase slowly to the aqueous phase while stirring constantly.
5. Remove the mixture from the water bath and stir until it congeals.

Cream type II: Water-soluble

1. Calculate the amount of ingredients you will need to prepare 10 g of water-soluble cream of composition 60 wt% polyethylene glycol 400 and 40 wt% polyethylene glycol 3350 (Carbowax 3350).
2. Melt the PEG 400 and Carbowax 3350 in a hot water bath around 65°C.
3. Remove the mixture from the water bath and stir until congealed. If you would like to add a fragrant to the cream (essential oil, e.g., lavender), you can mix the same with the other ingredients before the cream congeals, when the ointment is slightly warm.

Post-laboratory questions

1. Explain the role of each ingredient in your cream formulation.
2. Describe the appearance and consistency of each ointment base by applying it on your skin. How easy is it to spread the ointment?

Note for the instructor

There are several different formulations to prepare ointments. The recipes given here are very basic recipes. When preparing the ointments, cooling should be done slowly at room temperature to achieve the desired consistency. Some students may be encouraged to cool their mixtures using an ice bath, and the differences in the stiffness of the ointments may be compared to show the importance of method of preparation in obtaining the desired product. Ointments can also be prepared with an active agent in them, such as vitamins (oil-soluble) or salicylic acid (water-soluble, anti-acne treatment drug). The active ingredient should be delivered in the appropriate ingredient for best results.

References

Duman, G. *Pharmaceutical Technology Lab. 1: Dosage Forms and Contemporary Practice.* Faculty of Pharmacy, Yeditepe University, 2006. Experiment adapted with permission.

Monzer, F. *Surfactant Science: Colloids in Drug Delivery.* CRC Press, Boca Raton, FL, 2010.

Shrewbury, R.P. *The Pharmaceutics and Compounding Laboratory at University of North Carolina.* University of North Carolina, 2012. http://pharmlabs.unc.edu.

Thompson, J.E. *Contemporary Guide to Pharmacy Practice*, 3rd Edition. Lippincott Williams & Wilkins, Beijing, China, 2009.

Index